Occurrence of MTBE and VOCs in Drinking Water Sources of the United States

The mission of the Awwa Research Foundation (AwwaRF) is to advance the science of water to improve the quality of life. Funded primarily through annual subscription payments from over 1,000 utilities, consulting firms, and manufacturers in North America and abroad, AwwaRF sponsors research on all aspects of drinking water, including supply and resources, treatment, monitoring and analysis, distribution, management, and health effects.

From its headquarters in Denver, Colorado, the AwwaRF staff directs and supports the efforts of over 500 volunteers, who are the heart of the research program. These volunteers, serving on various boards and committees, use their expertise to select and monitor research studies to benefit the entire drinking water community.

Research findings are disseminated through a number of technology transfer activities, including research reports, conferences, videotape summaries, and periodicals.

Occurrence of MTBE and VOCs in Drinking Water Sources of the United States

Prepared by:
Bart Koch and **Melissa Dale**
Metropolitan Water District of Southern California
700 Moreno Avenue, La Verne, CA 91750

John Zogorski, **Greg Delzer**, **Stephen Grady**, **Tammy Ivahnenko**, and **Rick Clawges**
U.S. Geological Survey

Paul Tratnyek
Oregon Health & Science University

Sponsored by:
Awwa Research Foundation
6666 West Quincy Avenue, Denver, CO 80235-3098

Published by:

Awwa Research Foundation
Advancing the Science of Water®

IWA Publishing

DISCLAIMER

This study funded by the Awwa Research Foundation (AwwaRF). AwwaRF assumes no responsibility for the content of the research study reported in this publication or for the opinions or statements of fact expressed in the report. The mention of trade names for commercial products does not represent or imply the approval or endorsement of AwwaRF. This report is presented solely for informational purposes.

Published by IWA Publishing, Alliance House, 12 Caxton Street, London SW1H 0QS, UK

Telephone: +44 (0) 20 7654 5500; Fax: +44 (0) 20 7654 5555; Email: publications@iwap.co.uk;
Web: **www.iwapublishing.com**

AwwaRF report number 90954F. Originally published by AwwaRF for its subscribers in 2003.
IWA Publishing version published 2004.

ISBN 978-1-84339-860-8

TABLE OF CONTENTS

* The printed report ends on p. 190 and the Appendices on CD begin on p. 211. There are no pages missing.

LIST OF TABLES

LIST OF FIGURES

FOREWORD

The Awwa Research Foundation is a nonprofit corporation that is dedicated to the implementation of a research effort to help utilities respond to regulatory requirements and traditional high-priority concerns of the industry. The research agenda is developed through a process of consultation with subscribers and drinking water professionals. Under the umbrella of a Strategic Research Plan, the Research Advisory Council prioritizes the suggested projects based upon current and future needs, applicability, and past work; the recommendations are forwarded to the Board of Trustees for final selection. The foundation also sponsors research projects through the unsolicited proposal process; the Collaborative Research, Research Applications, and Tailored Collaborations programs; and various research efforts with organizations such as the U.S. Environmental Protection Agency (USEPA), the U.S. Bureau of Reclamation, and the Association of California Water Agencies.

This publication is a result of one of these sponsored studies, and it is hoped that its findings will be applied in communities throughout the world. The following report serves not only as a means of communicating the results of the water industry's centralized research program but also as a tool to enlist the further support of the nonmember utilities and individuals.

Projects are managed closely from their inception to the final report by the foundation's staff and large cadre of volunteers who willingly contribute their time and expertise. The foundation serves a planning and management function and awards contracts to other institutions such as water utilities, universities, and engineering firms. The funding for this research effort comes primarily form the Subscription Program, through which water utilities subscribe to the research program and make an annual payment proportionate to the volume of water they deliver and consultants and manufacturers subscribe based on their annual billings. The program offers a cost-effective and fair method for funding research in the public interest.

A broad spectrum of water issues is addressed by the foundation's research agenda: resources, treatment and operations, distribution and storage, water quality and analysis, toxicology, economics, and management. The ultimate purpose of the coordinated effort is to assist water suppliers to provide the highest possible quality of water economically and reliably. The true benefits are realized when the results are implemented at the utility level. The foundation's trustees are pleased to offer this publication as a contribution toward that end.

Edmund G. Archuleta, P.E.
Chair, Board of Trustees
Awwa Research Foundation

James F. Manwaring, P.E.
Executive Director
Awwa Research Foundation

ACKNOWLEDGMENTS

All chemical analyses for VOCs in source-water samples were conducted at the Metropolitan Water District of Southern California (MWDSC) Water Quality Laboratory in La Verne, California, under the direction of Bart Koch and Melissa Dale. The MWDSC chemists Lely Suhady and Eric Crofts are also thanked for their help in the analysis of the numerous samples. The authors would like to thank George D. Casey, Barbara Rowe, Brandon Taglio, and Erika Schoen of the U.S. Geological Survey for their efforts in compiling, analyzing, and interpreting all information provided during this project. The authors also thank Wentai Luo and Lorne Isabelle at the Oregon Health & Science University for performing the analysis of the degradation by-product samples.

The authors would like to thank the American Water Works Association Research Foundation (AwwaRF) and Kenan Ozekin for their support and oversight of this study. Also, the AwwaRF project advisory committee, consisting of Homer C. Emery of the San Antonio Water System, David Lipsky of the New York City Department of Environmental Protection, and Howard Weinberg of the University of North Carolina, are recognized for their technical advice. Importantly, this assessment would not have been possible without the cooperation and assistance of the owners, operators, and staff of the CWSs that conscientiously collected samples or allowed access to sample their source waters. The authors are also appreciative of CWSs for ancillary information provided on the characteristics of their water sources and the operation of their water system.

EXECUTIVE SUMMARY

BACKGROUND

The large-scale use of the gasoline oxygenate methyl *tert*-butyl ether (MTBE), in conjunction with its high solubility, low soil adsorption, and low biodegradability, has resulted in its detection in ground water and surface water in many places throughout the United States. Studies by numerous researchers, as well as many state and local environmental agencies, have discovered high levels of MTBE in soils and ground water at leaking underground gasoline-storage-tank sites and frequent occurrence of low to intermediate levels of MTBE in reservoirs used for both public water supply and recreational boating.

In response to these findings, the American Water Works Association Research Foundation sponsored an investigation of MTBE and other volatile organic compounds (VOCs) in the Nation's sources of drinking water. The Metropolitan Water District of Southern California (MWDSC) in collaboration with U.S. Geological Survey (USGS) and the Oregon Health & Science University (OHSU) has conducted a survey to study the frequency of occurrence, concentration, and distribution of MTBE, other ether gasoline oxygenates, their degradation products, and 62 other VOCs in source water used by community water systems (CWSs) in the United States.

DESIGN

The investigation was completed in two stages: (1) reviews of available literature and (2) the collection of new data. A review of literature conducted in 1997 for MTBE in water concluded that it was not possible to characterize MTBE in sources of drinking water due to the limited data available at that time. As such, literature reviewed for this report focused on those assessments completed after the 1997 review. Specifically, this literature review focused on public and private water-supply assessments that were national, regional, or statewide in scope. In total, 3 national, 2 regional, and 13 statewide assessments were reviewed. In addition, a review of literature on the aesthetic effects of MTBE was conducted.

Two surveys to collect new water quality data were conducted in this project. The first survey, termed the Random Survey, employed a statistically stratified design for sampling source water from randomly selected CWSs. The study design was statistically stratified for sampling CWSs by the total number of systems within five population-served size categories, source of water (ground or surface water), and the total number of people served by each of the source-size categories presently used within the United States. A total of 954 CWSs that represent the more than 54,000 CWSs from all 50 States, Native American Lands, and Puerto Rico were sampled. The 954 source-water samples included 579 samples from ground-water-supplied CWSs (wells, springs, and galleries) and 375 samples from surface-water-supplied CWSs (rivers, aqueducts, canals, lakes, and reservoirs). Samples were collected by CWS personnel and were analyzed by the MWDSC Water Quality Laboratory. Ancillary information about each participating CWS and their sources was collected to allow additional statistical analysis of the water-quality data.

The second survey, termed the Focused Survey, included samples collected from 134 CWS source waters, including ground water, reservoirs, lakes, rivers, and streams that were suspected or known to contain MTBE. The general intent of the Focused Survey was to compare results with the Random Survey, and to provide an improved understanding of the occurrence, concentration, temporal variability, and anthropogenic factors associated with primarily MTBE and also other frequently detected VOCs. CWSs located in hydrologic, climatic, and/or demographic settings that were suspected or known to be vulnerable to MTBE contamination were selected for sampling. CWS selection for the Focused Survey was based on CWSs that had source water suspected of having concentrations of MTBE. The initial batch of CWSs were chosed from CWSs located in 14 metropolitan areas with populations greater than 250,000 and where reformulated gasoline (RFG) was used. The subsequent selection of CWSs was based on the most recently completed or on-going sampling activities (including the Random Survey) and from water quality monitoring results available from individual states.

The 134 source waters were sampled at multiple frequencies depending on the type of source. Groundwater samples were collected bi-annually, reservoir/lake samples were collected quarterly and river/stream samples were collected 8 times per year. Therefore, a total of 451 source-water samples, which included 78 ground-water, 39 reservoir and/or lake, and 17 river and/or stream source waters, were collected at fixed intervals for a period of 1 year. All samples in the Focused Survey were collected by USGS personnel using USGS sampling protocols. Samples for the Focused Survey were analyzed by the MWDSC for MTBE, three other ether gasoline oxygenates, and 62 other VOCs. Duplicate samples were collected for about one-half of the Focused Survey samples and analyzed for oxygenate degradation by-products by OHSU. As with the Random Survey, ancillary information for the Focused Survey was collected to allow additional statistical analysis of the MTBE and other VOC data.

CONCLUSIONS

Literature Survey

The team's review of the literature found inconsistencies among the 18 assessments reviewed included different objectives, reporting levels, and different water types sampled (such as ambient, treated and untreated drinking water). This usually made comparisons among assessments difficult, if not impossible, to achieve. Regardless, results of this literature review indicated that MTBE has been detected in public and (or) private drinking-water supplies in 36 states at concentrations ranging from 0.1 to 17,800 μg/L (micrograms per liter). However, when median detected concentrations were available, they were typically low--less than 5.0 μg/L. In addition, the reviewed assessments collectively indicated that: (1) MTBE occurred in public drinking-water systems supplied by ground and surface water, and concentrations generally were less than 20 μg/L; (2) population density and reformulated gasoline use were significant risk factors for MTBE detection in water supplies; and (3) type of well, water supply, and proximity to gasoline storage tanks did not seem to be associated with MTBE detection.

A review of existing information on the aesthetic effect of MTBE in drinking water was also conducted. MTBE has an odor, which has been described as a sweet solvent (such as turpentine). Although a definitive taste and odor threshold level has not been well defined, the literature indicated that relatively low levels, ranging from 14 to 71 μg/L can cause an unpleasant aroma and flavor to the water. This range is congruous with the Environmental Protection Agency (EPA) advisory range of 20 to 40 μg/L. However, California has adopted a secondary standard (based on taste and odor) of 5 μg/L.

Random Survey

MTBE was detected in 8.7 percent of randomly selected CWS source waters in the United States at concentrations that ranged from 0.2 to 20 μg/L. At a reporting level of 0.2 μg/L, VOCs were detected in 27 percent of source-water samples collected from May 3, 1999 through October 23, 2000. Of 42 VOCs detected, chloroform (in 13 percent of samples) was detected most frequently in source-water samples, followed by MTBE. Chloroform in source waters may represent the presence of residual concentrations of disinfection by-products from upstream or upgradient discharges of chlorinated water or wastewater, an artifact of back-flushing water plant filters, industrial solvent sources, or sampling error. VOC concentrations were generally less than 10 μg/L—95 percent of 530 total detections—and 63 percent were less than 1.0 μg/L. Concentrations of 1,1-dichloroethene, tetrachloroethene, trichloroethene, vinyl chloride, and total trihalomethanes (THMs), however, exceeded drinking-water regulations in eight samples.

Detections of most VOCs were more frequent in surface-water sources than in ground-water sources, with gasoline compounds collectively, and MTBE individually, detected significantly more often in surface water. Use of personal and commercial motorized watercraft on surface-water bodies that are drinking-water sources is probably the reason for the elevated detections of gasoline contaminants relative to ground water. MTBE detections demonstrated a seasonal pattern with more frequent detections in surface water in summer months, which is consistent with seasonal watercraft use.

The detection frequency of most VOCs was significantly related to urban land use and population density. Detections of any VOC, non-trihalomethane compounds, gasoline compounds collectively, the specific gasoline compounds benzene, toluene, ethylbenzene, and xylenes (BTEX), MTBE, solvents, and refrigerants were significantly greater in areas with more than 60 percent urban land use and (or) population density greater than 1,000 people per square mile than in source waters from less urbanized or lower population-density areas. MTBE detections were five times more frequent in source waters from areas with high MTBE use than in source waters from low or no MTBE use, but, unlike other gasoline compounds, MTBE detections were not significantly related to the density of gasoline storage tanks near drinking-water sources.

Focused Survey

There were 37 of the 66 VOCs detected in both surveys when results were censored at the minimum reporting level (MRL) of 0.2 μg/L (2.0 μg/L for 2-butanone). However, VOCs, especially MTBE and other gasoline oxygenates, were detected more frequently in the Focused Survey than in the Random Survey. As expected, MTBE was detected in 55.5 percent of the CWSs sampled in the Focused Survey and in 8.7 percent of those sampled in the Random Survey. Little difference in occurrence, however, was observed for trihalomethanes (THMs), which were detected in 16.4 and 14.8 percent of Focused Survey and Random Survey CWSs, respectively. This may indicate a pervasive occurrence of THMs in several source-water types, regardless of CWS size or geographic location.

Using data at or above the method detection limit to assess temporal variability and anthropogenic factors associated with frequent detection of select VOCs, concentrations in the Focused Survey in ground water, reservoir, and river source waters were typically less than 1 μg/L. Also, at a 95-percent confidence interval, no statistically significant differences were observed in comparing concentrations in the first and second ground-water sample. A weak seasonal pattern was observed in samples collected from reservoirs and lakes where gasoline oxygenates and other gasoline compounds were detected more frequently during spring and summer, presumably a result of increased use of motorized watercraft during these seasons. In contrast, seasonal patterns were not observed in samples collected from rivers and streams. The lack of seasonal differences in river and stream source waters sampled may indicate a common and continuous source of contamination.

The most frequently detected VOC (MTBE) and the two most frequently occurring subgroups of VOCs (gasoline oxygenates and THMs) detected in CWS source waters were further characterized to evaluate some anthropogenic factors that may better explain their frequent occurrence. Gasoline oxygenates were detected in 73.9 percent of all CWSs sampled. The concentration of gasoline oxygenates was slightly correlated with watercraft use on reservoirs inside high-MTBE-use areas (r^2=0.3783) but not outside of these areas (r^2=0.0242). In general, the concentration of gasoline oxygenates increased as watercraft use increased. THMs were detected in 47.8 percent of the CWSs supplied by surface water. The frequent occurrence of THMs in reservoir source waters was determined to be an artifact of disinfection and the recycling of chlorinated water to these reservoirs. All CWSs with frequent occurrence of THMs served by a reservoir indicated that chlorine was added to waters for various reasons and that the chlorinated water was then released back to, or upstream of, the reservoir or lake that was sampled.

The highest levels of *tert*-butyl alcohol TBA were observed in well waters. There appeared to be no seasonal effect of TBA occurrence. There was a positive correlation of TBA occurrence with MTBE occurrence. It appears that observed concentrations of TBA are enriched relative to MTBE (based on typical amounts of TBA associated with oxygenated gasoline formulations), which could be due to degradation of existing MTBE, differential dissolution and transport of TBA relative to MTBE, and additional sources of TBA other than a degradation by-product of MTBE. There was no pattern of distribution of TBA relative to state or region.

Recommendations

Recommendations are discussed in chapter 7.

CHAPTER 1
INTRODUCTION

BACKGROUND

The 1990 Clean Air Act (CAA) Amendments mandated seasonal or year-round use of oxygenated compounds (oxygenates) in gasoline in specific parts of the United States. Oxygenates are added to gasoline to increase the oxygen content, which enhances combustion and decreases vehicular carbon monoxide emissions. Oxygenates also reduce the need for benzene and other ozone-forming, aromatic compounds in gasoline. Methyl *tert*-butyl ether (MTBE) is the most commonly used gasoline oxygenate in the United States, followed by ethanol. Oxygenates are added to gasoline during the winter months in areas where winter concentrations of carbon monoxide exceed established air-quality standards. This gasoline is called oxygenated (OXY) gasoline, and contains oxygenates at 2.7 percent by weight (15 percent by volume for MTBE). In select areas, oxygenates are added to gasoline year round to abate ozone pollution during the summer months as well as carbon monoxide pollution during the winter months (U.S. Environmental Protection Agency 1999a). This gasoline is called reformulated gasoline (RFG), and contains oxygenates at 2 percent by weight (11 percent by volume for MTBE). MTBE also has been added to gasoline as an octane enhancer since the late 1970's. With the phaseout of tetraethyl lead, MTBE has been used nationwide at low concentrations in conventional gasoline since 1979 to enhance octane levels.

MTBE production and use have increased substantially during the 1990's resulting from the implementation of the CAA Amendments. For example, MTBE went from the 39th highest produced organic chemical in the United States in 1970 to the fourth highest in 1998 (Johnson et al. 2000). During that period, the aggregate production of MTBE was 60 million metric tons. In 1998, 10.5 million gallons (39.7 million liters) per day were used in the United States, 40 percent of which was used in California alone (Johnson et al. 2000). The majority of MTBE use is associated with RFG. MTBE is used in only about 3 percent of OXY gasoline, whereas MTBE is used in about 85 percent of all RFG (Wigglesworth 1999). As of 2000, 10 areas in 9 States are involved in an OXY gasoline program, and 11 areas in 17 States are involved in an RFG program (Figure 1.1).

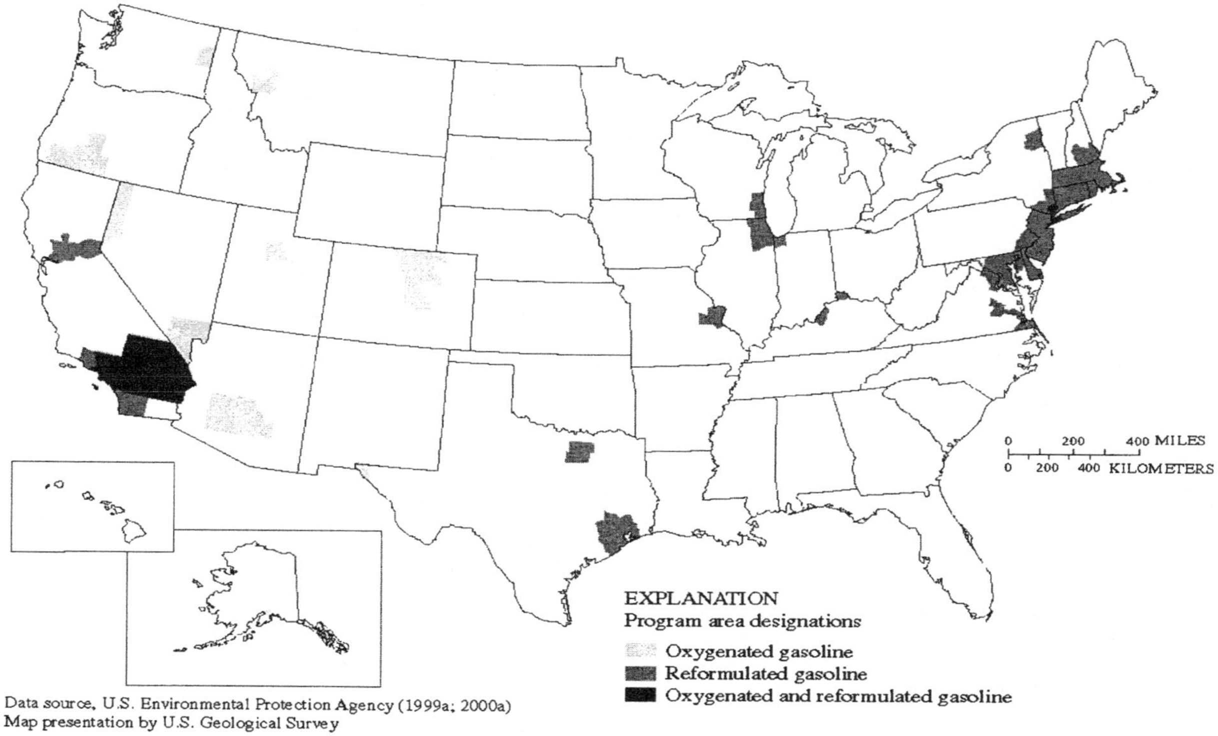

Figure 1.1 Location of current (2000) Federal Oxygenated and Reformulated Gasoline Program Areas

Figure 1.1 Location of current (2000) Federal oxygenated and reformulated gasoline program areas

In 1996, the U.S. Environmental Protection Agency (USEPA) had tentatively classified MTBE as a Group C, possible human carcinogen (U.S. Environmental Protection Agency 1996). However, in 1997, the USEPA issued a drinking water advisory that recommends a range of 20 to 40 μg/L as a limit to avert unpleasant taste and odor effects (U.S. Environmental Protection Agency 1997). The advisory concentration range is intended to provide a safety margin (4 to 5 orders of magnitude) for potential carcinogenic effects and to provide a large margin of safety for non-cancerous effects (U.S. Environmental Protection Agency 1997). The State of California has issued a taste and odor limit of 5 μg/L and a maximum contamination level (MCL) of 13 μg/L (California Department of Health Services 2001). Currently (2002), MTBE is on the Candidate Contaminant List. The USEPA has required that monitoring for MTBE be conducted by selected community water systems (CWSs) under the Unregulated Contaminant Monitoring Rule (U.S. Environmental Protection Agency 1999c). More health effects data are required for this contaminant before the USEPA can determine whether to regulate it (U.S. Environmental Protection Agency 1998a). A national MCL for MTBE has not yet been issued by the USEPA.

MTBE OCCURRENCE

The chemical characteristics of MTBE, such as high solubility in water, low Henry's Law constant, low soil-sorption properties, and recalcitrant nature in ground water, may cause MTBE contamination of public and private drinking-water sources. MTBE may be released into ground water and surface water from point sources, such as leaks or spills, especially during the distribution, storage, and use of blended gasoline (Zogorski et al. 1997; Hitzig, Kostecki, and Leonard 1998; Hunter 1999; Moran et al. 1999a). MTBE may also be released from nonpoint sources, such as automobile and watercraft emissions and evaporative losses, urban precipitation and stormwater runoff (Pankow et al. 1997; Baehr, Stackelberg, and Baker 1999; Moran et al. 1999a). MTBE has also been found in spills of home heating oil in northeastern states, where it may occur as a contaminant in the fuel oil as a result of mixing with residual amounts of gasoline containing MTBE in the bulk storage, distribution, or delivery of the oil (Robbins et al. 1999).

The combination of MTBE's widespread use and frequent release to the environment through leaks and spills, together with the aforementioned chemical characteristics has caused extensive contamination of private and public drinking-water supplies. Some cities, especially in California, have already lost a substantial number of drinking-water sources. For example: in Santa Monica, 75 percent of the drinking-water wells are unusable due to MTBE; in South Lake Tahoe, California, one-third of the city's 34 drinking-water wells have been shut down because of MTBE contamination; in Windam, Maine, wells have been taken offline; and Los Angeles, San Francisco, Santa Clara Valley, and Sacramento (in California) all have wells affected by MTBE (Bourelle 1998; City of Santa Monica 1999; State of Maine 1998; California Department of Health Services 2001). La Crosse, Kansas, also has had wells affected by MTBE; however, officials there have taken steps to remediate the problem and to treat the water rather than to remove the wells from service (Hattan 2000).

Although isolated instances of MTBE contamination have been observed, the overall extent of MTBE occurrence in the nation's drinking-water supplies has not been evaluated. The Interagency Assessment of Oxygenated Fuels (Zogorski et al. 1997) attempted to address the national occurrence of MTBE in drinking-water supplies, but was unable to do so due to insufficient data. As a result, additional assessments were recommended.

MTBE DEGRADATION BY-PRODUCTS

Gasoline oxygenates are degraded in the environment by a variety of chemical and biological processes. The former are mostly photochemical reactions occurring in the atmosphere and the latter are mostly microbiological reactions occurring in soils, sediments, and aquifers. In addition to these transformation processes, there are transport processes that distribute the oxygenates and their degradation by-products among environmental compartments (air, soil, surface water, ground water). The major transformation processes involved in the fate of MTBE are summarized in Figure 1.2.

The situation represented by Figure 1.2 suggests a variety of mechanisms by which drinking water sources might become contaminated with by-products of oxygenate degradation. The range of characteristic by-products that have been identified for biodegradation of gasoline oxygenates is represented in Figure 1.3. Although this scheme was developed from data for a pure culture of a particular MTBE-degrading bacterium (Hanson, Ackerman, and Scow 1999) its major features are consistent with most reported biodegradation pathways for gasoline oxygenates (Salanitro et al. 1994; Hardison et al. 1997; Steffan et al. 1997), and even some (photo) chemical oxidation pathways (Barreto et al. 1995; Kang and Hoffmann 1998). Analogous transformations can be expected for other alkyl ethers used as gasoline oxygenates (e.g., ethyl *tertiary* butyl ether [ETBE] and *tertiary* amyl methyl ether [TAME]). Thus, Figure 1.3 provides a general guide to the by-products of gasoline oxygenate degradation that might be found in drinking water supplies.

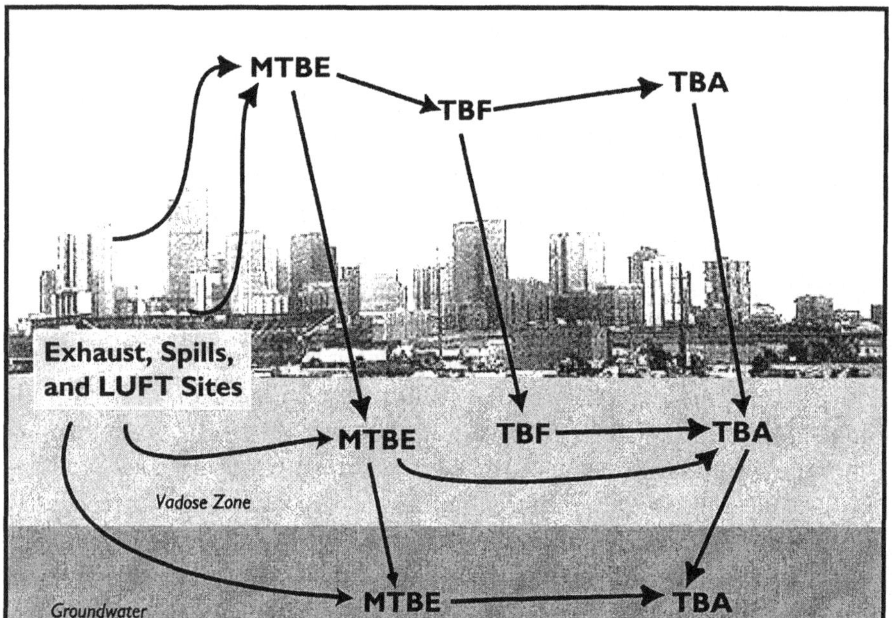

LUFT – leaking underground fuel tanks

Note: TBF (*tert*-butyl formate) is subject to hydrolysis to TBA (*tert*-butyl alcohol). TBA is subject to further biodegradation to acetone, isopropanol, formate, and ultimately CO_2. Adapted from (Church et al. 1999).

Figure 1.2 Summary of chemical transformation pathways and phase transfer processes for MTBE

Of the by-products shown in Figure 1.3, tert-butyl alcohol (TBA) is the most likely to accumulate to detectable levels in the environment from degradation of gasoline oxygenates. This is because TBA arises from degradation of MTBE and biodegradation of TBA is usually slow, comparable to the rate of biodegradation of MTBE. Tert-butyl formate (TBF) may arise as an intermediate in the degradation of MTBE to TBA, but hydrolysis of TBF to TBA is comparatively fast, so it is likely to be a less persistent (and therefore less diagnostic) by-product than TBA (Church, Pankow, and Tratnyek 1999). Although MTBE is also subject to hydrolysis (O'Reilly et al. 2001), this reaction is much slower than the hydrolysis of TBF and therefore is likely to be of much less environmental significance. Similar factors may govern the relationship between tert-amyl methyl ether (TAME), tert-amyl formate (TAF), and tert-amyl alcohol (TAA), but not enough work has been done on these compounds to be sure. The degradation products of TBA (isopropanol, etc.) are likely to further degrade, so they are unlikely to accumulate and therefore they are not promising candidates for by-products that are diagnostic of oxygenate degradation.

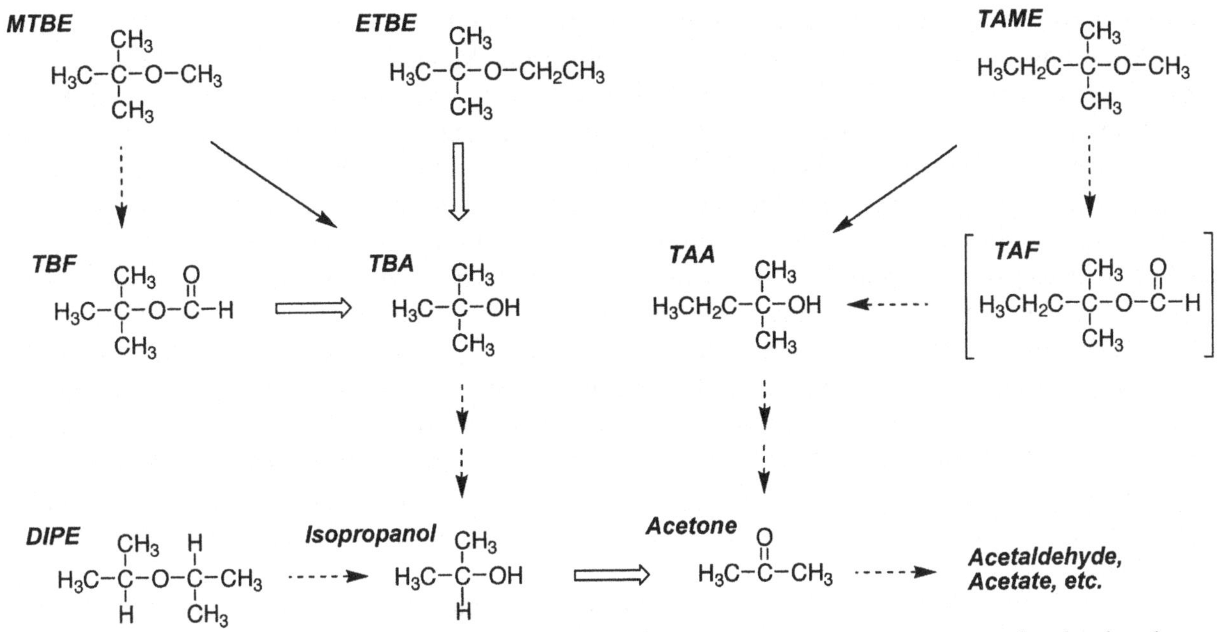

* Similar pathways of biodegradation have been reported for other pure and mixed cultures (Church et al., 2000).

Note: Open arrows: transformations that were directly observed. Solid arrows: observed but may occur via intermediates. Dashed arrows: hypothesized steps.

Figure 1.3 . Proposed degradation pathways* of fuel oxygenates by the MTBE degrading bacterium strain PM1.

When TBA is detected, it may be a by-product in the degradation of MTBE, but two other factors complicate this interpretation of TBA occurrence data. First, TBA has been used as an oxygenate in some areas, and TBA is present as an impurity in much of the MTBE used to oxygenate gasoline. Second, the dissolution of TBA from spilled gasoline, and the transport of TBA in the groundwater may be more rapid than for MTBE because the former has a higher water solubility and lower sorption coefficient (Church et al. 1999; Squillace et al. 1997). These complications have limited the utility of TBA occurrence as evidence for MTBE degradation in the field to a few intensively studied sites (Landmeyer et al. 1998; Salanitro et al. 2000).

STUDY DESIGN

Since MTBE has been detected in some surface and ground waters, the need exists to (1) expand the occurrence database on MTBE, other ether oxygenates, and MTBE degradation by-products and (2) ensure adequate inclusion of both ground and surface drinking water supplies in the study design. Previous studies have focused on the northern and mid Atlantic states, and primarily on groundwater. This project was proposed and funded by the American Water Works Association Research Foundation (AwwaRF) to help provide a comprehensive and representative sampling of MTBE occurrence in ground and surface water for the entire nation. The Metropolitan Water District of Southern California (MWDSC) in La Verne, California, in collaboration with the U.S. Geological Survey (USGS) National Water Quality Assessment (NAWQA) Program's National Volatile Organic Compound (VOC) Synthesis Team, and the Oregon Health & Science University (OHSU) in Beaverton, Oregon, has designed and conducted a nationwide survey to determine the occurrence and distribution of MTBE, MTBE degradation by-products, other ether oxygenates, and other VOCs in ground and surface water that serve as public drinking-water supplies. A wide range of VOCs was included to place MTBE occurrences in context with other similar compounds. This survey provided ,a basis for determining the extent of MTBE and other VOC contamination throughout the United States by sampling a large enough population of CWS source waters to allow statistical analysis with a high degree of confidence. In addition, the survey provided information about the frequency of detection, concentration, and temporal variability of MTBE and other gasoline oxygenates, their degradation products, and other VOCs in source water with suspected or known MTBE contamination. This survey also included the analysis of TBA, the major degradation by-product of MTBE.

The information from this survey will help accomplish the goals of the USGS VOC National Synthesis Team by determining the occurrence and distribution of other VOCs in ground and surface water that serve as public drinking-water supplies. The complete list of compounds included in this assessment also included 14 VOCs that are on the USEPA's Contaminant Candidate List (CCL) (U.S. Environmental Protection Agency 1998a). The CCL is a list of contaminants not currently regulated by a National Primary Drinking Water Standard which may be regulated in the future. Additional data on these compounds are needed before a regulatory determination can be made. This national assessment of MTBE, MTBE degradation by-products, other ether oxygenates, and other VOCs began in September 1998.

Drinking-water supplies studied were specifically termed source waters. Source water is defined by the American Water Works Association as "the supply of water for a water utility [that] is usually treated before distribution to consumers" (Symons, Bradley, and Cleveland 2000). A source water can be a river, brook, stream, lake, reservoir, impoundment, spring, or aquifer from which a supply of water is obtained. As such, the assessment did not study any waters that were within a distribution line or after treatment processes have been applied to the water.

The national assessment of MTBE and other VOCs in sources of public drinking-water supplies has been accomplished by a two stage approach: (1) reviews of available literature, and (2) the collection of new drinking-water-quality data. There are two literature reviews associated with this assessment. One review focused on MTBE taste and odor threshold concentrations and their relevance to aesthetic effects and possible water-treatment requirements. The second review concentrated on the occurrence of MTBE in public and private drinking-water supplies reported by national, regional, or statewide assessments.

Two surveys were associated with the collection of new drinking-water data. The first survey, termed the Random Source-Water Survey (hereafter termed Random Survey), employed a statistically stratified design for sampling source water from 954 community water systems (CWSs) randomly selected on the basis of the total number of systems within five population-served size categories, source-water type, and the total number of people served by each of the source-size categories used within the United States. The Random Survey was designed to determine the frequency of detection and the range in concentrations of MTBE and other VOCs in drinking-water sources. The 954 source-water samples were distributed to allocate 579 samples to ground-water-supplied CWSs (wells, springs, and galleries) and the remaining 375 samples to surface-water-supplied CWSs (rivers, aqueducts, canals, lakes, and reservoirs). Sampling for the Random Survey was begun on May 3, 1999 and was completed on October 23, 2000. Samples were collected by CWS staff and analyzed by the MWDSC for MTBE, three other ether gasoline oxygenates, and 62 other VOCs using U.S. Environmental Protection Agency Method 524.2. Ancillary information about the participating CWSs and their sources was collected to allow additional statistical analysis of the water-quality data.

The second survey, termed the Focused Source-Water Survey, (hereafter termed Focused Survey) was based on factors that appear or are known to be related to frequent detection of MTBE and other VOCs in source water. The Focused Survey provided an improved understanding of the occurrence, concentration, temporal variability, and anthropogenic factors associated with the detection of MTBE, TBA, ether oxygenates, and other frequently identified VOCs. CWSs located in hydrologic, climatic, and/or demographic settings that were suspected or known to be vulnerable to MTBE contamination were selected for sampling.

A total of 451 samples were collected in the Focused Survey, from 78 ground water sources, 80 wells, 39 reservoirs/lakes, and 17 rivers/streams across the United States. Sampling was begun on August 18, 1999 and was completed on May 25, 2001. Samples for the Focused Survey were collected by USGS personnel and analyzed by the MWDSC for MTBE, three other ether gasoline oxygenates, and 62 other VOCs using U.S. Environmental Protection Agency approved Method 524.2. Additional samples collected from about one-half of these were also analyzed for oxygenate degradation products by OHSU. As with the Random Survey, ancillary information for the Focused Survey was collected to allow additional statistical analysis of MTBE and other VOC data.

CHAPTER 2
LITERATURE SURVEY

REVIEW OF LITERATURE FOR METHYL *TERT*-BUTYL ETHER IN SOURCES OF DRINKING WATER IN THE UNITED STATES

Purpose and Scope

The literature review presented in this report provides information on the frequency of detection, concentration, and distribution of MTBE in sources of water used for drinking water and in finished drinking water in the United States. The review focused on public and private water-supply assessments that were national, regional, or statewide in scope. Specifically, 3 national, 2 regional, and 13 statewide assessments were reviewed. In some cases, results are reported from summaries of information provided by states and are based on the state's normal compliance monitoring rather than a designated assessment. As of 2001, no additional large-scale MTBE drinking-water assessments were known to exist. Large-scale assessments were targeted for this literature review to provide a basis for comparison to the results obtained from the second phase of the national AwwaRF assessment—the collection of new drinking-water-quality data.

Summaries of MTBE in Drinking Water

The following summaries were obtained from assessments completed that were national, regional, or statewide in scope. The first national assessment, an Interagency Assessment of Oxygenated Fuels (Zogorski et al. 1997), reviewed all available scientific literature, data, and agency information with respect to all fuel oxygenates and their degradation by-products as of 1997. As such, this report serves as a comprehensive compilation of available information on ether oxygenates in drinking water up to the time of its completion. Additional assessments described herein were completed after the Interagency Assessment of Oxygenated Fuels report. An overall comparison of each assessment's findings is presented in Table 2.1 in the "Summary" section of this chapter. However, direct comparisons of findings among assessments generally were not possible due, in part, to variable reporting limits and different types of water sampled (untreated drinking water, finished drinking water, and ambient water).

National Assessments

Three national assessments are summarized in this section, including the previously mentioned Interagency Assessment of Oxygenated Fuels. The other two national assessments that are summarized include an assessment of MTBE in rural domestic wells and an assessment of MTBE identified by state Leaking Underground Storage Tank (LUST) programs.

Interagency Assessment of Oxygenated Fuels

In June 1997, the Executive Office of the President's Office of Science and Technology Policy released a report summarizing an Interagency Assessment of Oxygenated Fuels (Zogorski et al 1997). The purposes of this report were to provide a review of the scientific literature on oxygenated fuels and to assess the effects of the winter OXY gasoline program on air quality, water quality, fuel economy, engine performance, and public health.

As part of this assessment, two data sets were available to provide information on the occurrence of MTBE in drinking water. The first data set was collected by the USGS, mostly as part of the NAWQA Program (Gilliom, Alley, and Gurtz 1995), where MTBE was included on a list of analytes for ground-water samples collected in 20 study areas across the country during 1993-94. In addition, retrospective efforts of the NAWQA Program summarized MTBE occurrence data from a few state and regional ground-water assessment programs. The USGS data were a mixture of ambient ground-water data and drinking-water data. This is largely due to the NAWQA Program emphasis on resource assessments rather than characterizations of drinking water at that time.

The second data set was assembled while the Interagency Assessment of Oxygenated Fuels report was being prepared. The USEPA requested, through its 10 regional offices, information on drinking-water programs that have analyzed for MTBE. In response to this request, data were provided by seven states on the occurrence of MTBE in public drinking-water supplies derived from ground water, and four states provided data collected from domestic wells.

The USGS data indicated that MTBE was analyzed in samples collected from 1,516 wells and (or) drinking-water systems in 33 states with varying land uses. At least one detection of MTBE occurred in ground water in 14 of the 33 states surveyed. Overall, MTBE was detected in 5 percent (76 of 1,516) of the samples with a reporting level ranging from 0.2 to 1.0 μg/L. Most of the detections occurred in shallow ground water in urban areas (55 of 204 samples), whereas only about 1.3 percent (7 of 524) of the samples from wells in agricultural areas had detections of MTBE. Of the 76 total detections of MTBE, 13 occurred in drinking-water supplies in six states—Colorado, Connecticut, Georgia, New York, Pennsylvania, and Washington. Eleven of these 13 detections occurred in domestic water supplies, and two occurred in public water supplies. Concentrations detected in ground water and drinking water obtained from ground water (combined) ranged from 0.2 to 23,000 μg/L with a median of 0.6 μg/L. Concentrations detected in drinking water (domestic and public) generally were low and ranged from 0.2 to 2.2 μg/L with a median of 0.5 μg/L.

The USEPA request to states resulted in information on MTBE occurrence in public water systems voluntarily provided by seven states—Colorado, Illinois, Iowa, New Jersey, Rhode Island, Texas, and Wisconsin. Indiana, Missouri, Rhode Island, and Texas also provided information on domestic wells. The information provided was mostly specific to MTBE occurrence in public drinking-water supplies derived from ground water. However, Rhode Island and Wisconsin reported MTBE data for surface water as a source of drinking water.

Findings from the USEPA request indicated that MTBE was detected at least once in public water systems supplied from ground water in six of the seven states. MTBE was detected in 51 public drinking-water systems in Colorado, Iowa, Illinois, New Jersey, Rhode Island, and Texas. Of the 51 public drinking water systems with MTBE detections, 35 (68.8 percent, representing approximately 1,300 samples) were in New Jersey, whereas seven were in each - Illinois and Texas (13.7 percent representing an unknown number of samples and 15,352 samples, respectively), one was in Colorado, and one was in Iowa (2 percent representing five samples and an unknown number of samples, respectively) (Zogorski et al. 1997). When detected, the concentrations of MTBE generally were low and almost always less than 20 μg/L. However, concentrations as high as 63 μg/L in Iowa and 770 μg/L in Illinois were reported. In addition, MTBE was detected in all four states that provided data for domestic wells: Indiana,

Missouri, Rhode Island, and Texas. Samples from five domestic wells from Indiana and Missouri had MTBE concentrations exceeding 200 µg/L, and one had a concentration of 40 µg/L. It is important to note, however, that domestic well samples were collected in response to consumer complaints. Thus, the MTBE concentrations in domestic wells generally were high in comparison to public water-supply information. In general, the public water-supply findings also may be biased toward high concentrations because a number of sampling programs were conducted when contamination from a nearby point source was known or suspected.

In addition to the information provided by states on MTBE occurrence in drinking water derived from ground water, Rhode Island and Wisconsin provided data on drinking water derived from surface water. In Rhode Island, an unknown number of surface-water samples were analyzed for MTBE, and one detection of MTBE occurred at a system that used surface water as a source. The detection occurred in a sample collected in January 1994 with a concentration of 1 µg/L. Two drinking-water systems were sampled during April 1995 in Wisconsin, and MTBE was not detected.

Assessment of MTBE in Rural, Domestic Wells

The results of an assessment to characterize the occurrence, distribution, and levels of 55 VOCs in untreated, self-supplied water from rural, domestic wells in the United States were reported by Moran et al. (2002b). Wells were considered to be in rural areas if the population density around the well was less than 386 people per square kilometer. One of the 55 VOCs assessed was MTBE. Data used in this assessment were compiled from two sources. The first source was a compilation of VOC data collected from wells by the NAWQA Program during 1993-99. Some of these wells also were included in Zogorski et al. (1997). The second source was a compilation of VOC data collected and analyzed as part of ambient ground-water or source-water-quality monitoring programs by local, state, and other federal agencies during 1985-95. All wells included in these analyses were at least 1 kilometer away from each other to avoid overlapping data, and only one sample from each well was included.

MTBE data from a total of 1,335 domestic wells located in 39 states were evaluated. At a reporting level of 0.2 µg/L, MTBE was detected in 2.2 percent of samples (30 of 1,335) in eight states. Typically, detections were associated with areas where the RFG program was federally mandated and in areas near larger cities that opted for voluntary participation in the RFG program. The states in which MTBE was detected in domestic wells included: Arkansas (8.3 percent, 1 of 12 wells), Colorado (5.6 percent, 3 of 54 wells), Connecticut (25 percent, 3 of 12 wells), Georgia (9 percent, 2 of 22 wells), Illinois (1.9 percent, 1 of 52 wells), Massachusetts (33 percent, 2 of 6 wells), New Jersey (16.4 percent, 12 of 73 wells), and Pennsylvania (3.7 percent, 6 of 164 wells). Concentrations ranged from 0.2 to 30.2 µg/L with a median of 0.7 µg/L.

Summary of State Leaking Underground Storage Tank Programs

The University of Massachusetts, with support from the USEPA, Office of Underground Storage Tanks, developed a mail questionnaire to characterize the effect of MTBE on state LUST programs and to identify and evaluate any effective methods for dealing with MTBE that states have developed (Hitzig, Kostecki, and Leonard 1998). State LUST programs were surveyed, allowing respondents to choose very general responses to increase the response rate for the survey. Questions included whether or not MTBE was required to be analyzed and, if so, how frequently it was detected at LUST sites. However, the generalized responses limited the ability to obtain specific data. As such, these data are not presented in Table 2.1 in the "Summary" section of this report. Responses from LUST programs came from 48 states and the District of Columbia. Only California and Indiana did not respond.

In addition to the LUST site questions, programs also were asked to estimate the number of MTBE-contaminated drinking-water wells reported in their state or territory. Of all the LUST programs, 25 had reports of private drinking-water wells contaminated with MTBE. The total number of private wells contaminated with MTBE was estimated to range from 2,256 to 2,663. The largest number of contaminated domestic wells occurred in New York and was estimated to be at least 546 on the basis of responses from only four of New York's nine regions (Robert Hitzig, U.S. Environmental Protection Agency, Office of Underground Storage Tanks, written commun., 1998). States which reported MTBE contamination in 1 to 10 private drinking-water wells included Alabama, Idaho, Massachusetts, Minnesota, Montana, Oregon, and South Carolina. States which reported MTBE contamination in 11 to 20 private drinking-water wells included Delaware, Florida, North Carolina, and West Virginia. No states reported MTBE contamination in the 21- to 30-well range; however, Michigan and Wisconsin reported contamination in the 31- to 40-well range. Connecticut, Kansas, Maine, Maryland, New Hampshire, New Jersey, New Mexico, New York, Rhode Island, Tennessee, Vermont, and Virginia each reported contamination in more than 40 private drinking-water wells.

Nineteen LUST programs identified public drinking-water wells contaminated with MTBE. The total number of contaminated public wells was estimated to range from 251 to 422. States which reported estimated MTBE contamination in 1 to 10 public drinking-water wells included Connecticut, Florida, Michigan, Missouri, North Carolina, Rhode Island, South Carolina, Virginia, and West Virginia. States which reported estimated contamination in 11 to 20 public drinking-water wells included Maine, Maryland, New York, and Wisconsin. Kansas, Massachusetts, and New Mexico estimated 21 to 30 contaminated public drinking-water wells, and New Hampshire and Vermont reported between 31 and 40 contaminated public drinking-water wells. New Jersey estimated 65 contaminated public drinking-water wells (Robert Hitzig, U.S. Environmental Protection Agency, Office of Underground Storage Tanks, written commun., 1998).

Very limited data on MTBE concentrations were provided; however, when available, the vast majority of concentrations in public drinking-water wells were less than 10 μg/L. This assessment indicated that MTBE detections by state programs were common and that MTBE contamination may occur at diesel fuel storage sites or unexpected locations due, in part, to surface dumping of small amounts of gasoline.

Regional Assessments

Two regional assessments were reviewed and are summarized in this section. These assessments include a survey of MTBE in sources of drinking water by the American Water Works Company and an assessment of MTBE in finished drinking water from community water supplies in the Northeast and Mid-Atlantic regions of the United States.

American Water Works Company Assessment of MTBE in Drinking Water

Gullick and LeChevallier (2000) summarized the results from a MTBE survey of ground- and surface-water supplies used for drinking water in the American Water System of the

American Water Works Company. The system is large and comprises wholly owned utility subsidiaries in 23 states that serve more than 7 million people in 879 communities. A total of 1,349 ground-water samples representing 342 wells from 17 states were analyzed from 1997 to 1998 using a reporting level of 0.5 μg/L. Raw water was sampled from 270 of these wells, and treated water was sampled from 72 wells.

MTBE was detected in wells in eight states at a frequency of 8.8 percent (30 of 342). The highest concentration detected was 14.1 μg/L. The states in which MTBE was detected included Connecticut (50 percent, 5 of 10 wells), Indiana (1 percent, 1 of 99 wells), Maryland (50 percent, 1 of 2 wells), Massachusetts (38 percent, 5 of 13 wells), New Hampshire (9.1 percent, 1 of 11 wells), New Jersey (13 percent, 12 of 92 wells), Pennsylvania (100 percent, 4 of 4 wells), and West Virginia (50 percent, 1 of 2 wells). MTBE was detected in treated water in Massachusetts (50 percent, 2 of 4 samples) and New Jersey (9.1 percent, 4 of 44 treated water samples).

A total of 200 surface-water samples representing 92 sampling sites in 12 states also were analyzed. MTBE was detected at 8 of 92 sites (8.7 percent) in three states. The highest concentration detected was 25.1 μg/L. States in which MTBE was detected in surface water included New Jersey (25 percent, 4 of 16 sites), New York (17 percent, 3 of 18 sites), and Pennsylvania (11 percent, 1 of 9 sites).

Characterization of MTBE in Finished Drinking Water in the Northeast and Mid-Atlantic Regions

Grady and Casey (2001) characterized the occurrence and distribution of MTBE and other VOCs in finished drinking water from CWSs in the Northeast and Mid-Atlantic regions of the United States from 1993 to 1998. These regions were selected for assessment because of their large populations, extensive urban and industrial development, and widespread use and release of many VOCs. In addition, these regions comprise the largest contiguous area, outside of California, where the gasoline additive MTBE is used to meet requirements of the CAA Amendments of 1990. The assessment area included the six New England States plus Delaware, Maryland, New Jersey, New York, Pennsylvania, and Virginia.

This assessment was designed to provide statistically representative data for each of the 12 states. A random selection of 20 percent (2,110) of the 10,479 active CWSs in the assessment area (as of December 1, 1997) was made to represent the actual distribution of CWSs in the area. The resulting distribution of randomly selected CWSs included 1,690 systems supplied exclusively by ground water, 270 supplied exclusively by surface water, and 150 systems with both ground- and surface-water sources. About 65 percent of the selected CWSs were small, serving fewer than 500 people, 20 percent served 501 to 3,300 people, 13 percent served 3,301 to 50,000 people, and 2 percent served more than 50,000 people. The number of systems selected from within each state was proportional to the size of the population within that state and ranged from 30 CWSs in Rhode Island to 538 CWSs in New York.

Chemical analyses, as well as supporting documentation, were provided for each of the randomly selected CWSs by each respective state. Data obtained from states varied significantly in format, constituents analyzed, reporting levels, and period of record. As such, MTBE was not analyzed at each of the randomly selected CWSs; a total of 5,510 analyses of MTBE from 1,194 CWSs were provided. MTBE data were not available from Delaware or Pennsylvania, however. Reporting levels ranged from 0.5 to 10.0 μg/ L. At any reporting level, MTBE was found to occur in 6.2 percent (343 of 5,510 analyses) of the samples and 8.9 percent (106 of 1,194) of the CWSs. Further examination of these data by applying censoring levels of 20, 10, 5, 2, 1, and 0.5 μg/L resulted in frequencies of detection of 0.8, 1.4, 2.0, 4.6, 7.8, and 12.2 percent, respectively. It should be noted that 4,427 of the 5,510 analyses at 865 of the 1,194 CWSs were included using a reporting level of 0.5 μg/L. Thus, only about 20 percent of the data set had a reporting level higher than 0.5 μg/L.

MTBE was detected in drinking water from all 10 states in the two regions that had analytical data. MTBE concentrations ranged from 0.26 to 210 μg/L; however, most MTBE concentrations were less than 5.0 μg/L, and 0.8 percent of CWSs reported MTBE concentrations equal to or greater than 20 μg/L. Two percent of the CWSs reported MTBE concentrations equal to or greater than 5 μg/L.

Although MTBE concentration data were not available on a state-by-state basis in the report by Grady and Casey (2001), the frequency of detection in randomly selected CWSs varied substantially from state to state at a reporting level of 1.0 μg/L. MTBE was detected most frequently in New Jersey (about 22 percent) and least frequently in Virginia (1.3 percent). MTBE was found to occur in Connecticut and Rhode Island approximately 17 percent of the time. MTBE also was found to occur at frequencies of approximately 12, 11, 10, 9, 4, and 2 percent in New Hampshire, Massachusetts, Maine, New York, Vermont, and Maryland, respectively. The states with the highest MTBE detection frequencies— New Jersey, Rhode Island, and Connecticut—are the three states with the highest population density. MTBE occurrence was shown to be statistically related to population density using a contingency-table test (p-value less than 0.0001).

Statewide Assessments

Statewide assessments of MTBE have been conducted in 13 states by various entities and are summarized in this section. The reporting level, if given, varied between studies as did the types of water sampled (sources of drinking water, finished drinking water, and ambient water). Certain states (including in part, California, New Jersey, New Hampshire, and New York) have required monitoring and assembled data sets and set state MTBE MCLs.

Alabama

Due to concern of MTBE contamination and based in part on the results of the national and regional assessments just discussed, the Alabama Department of Environmental Management (ADEM), Water Division (2001) conducted an assessment to determine if the sources of water used by public water-supply (PWS) systems in Alabama have been affected by MTBE. MTBE has been and continues to be present in gasoline sold in Alabama at levels typically between 1 to 2 percent and occasionally as high as 4 percent by volume.

There are 575 PWS systems in Alabama that use surface water, springs, and wells to meet water-supply demands. Sixty percent of all the water supplied comes from surface water. In calendar year 2000, each of the 575 PWS systems was sampled for MTBE. Efforts were made to sample each potable water source prior to treatment at each PWS system. As a result, 1,053 water sources were sampled during the study, including 87 surface-water sources, 27 springs, and 939 wells.

Most surface-water sources were sampled four times; once during March to April, once during May to June, once within 3 days following the Fourth of July, and once within 4 days of Labor Day. This sampling strategy accounted for some temporal fluctuations of MTBE in surface water due to lake turnover, rainfall, runoff, and periods of high recreational use. The wells and springs were sampled once, however MTBE will be analyzed on a routine basis in the future, as it is currently on the standard state VOC analyte list. Samples were analyzed by the Tennessee Valley Authority or an ADEM-approved certified laboratory. As such, reporting levels ranged from 0.5 to 2.0 μg/L depending on the laboratory used.

MTBE was not detected in any of the 87 surface-water sources. Five of 939 wells (0.53 percent) contained MTBE with concentrations ranging from 0.74 to 8.39 μg/L. Twenty-seven springs were sampled and analyzed for MTBE; all results were below the detection limit. ADEM staff are currently (2001) conducting follow-up investigations to try to determine the source of MTBE in the five contaminated wells.

California

California State Senate Bill 521, which became effective January 1, 1998, called for the University of California to perform an assessment of the benefits and risks associated with the uses of MTBE in California. The assessment reported by Keller et al. (1998) addressed, in part, current effects of MTBE in drinking-water sources from surface water and ground water. Data were obtained from the Department of Health Services, Local Primacy Agencies, and Regional Water Quality Control Boards within California.

A total of 245 surface-water sources, including reservoirs, lakes, and rivers, were identified as drinking-water sources. Data were available for 105 of these surface-water sources. About 3,000 samples were collected and analyzed voluntarily during 1996-97; however, the vast majority of samples were from 25 of the 105 surface-water sources and were collected on multiple days at several different locations within each water body. MTBE was not detected in 56 (53 percent) of the 105 water bodies sampled; 49 (46.7 percent) of the 105 water bodies did contain MTBE. Twenty-six (53 percent) of these 49 water bodies were found to have MTBE concentrations greater than 5 μg/L, and 13 (26 percent) of the 49 water bodies were found to have MTBE concentrations greater than 14 μg/L in at least one sample.

Of 13,919 public-supply wells within California, 2,988 were reported to have been sampled and analyzed for MTBE. Of these 2,988 wells, MTBE contamination was reported in 35 wells. This equates to 1.2 percent of all public supply wells tested for MTBE and 0.3 percent of all public-supply wells in counties where at least one well was tested. Through extrapolation of these data using a 1.2 percent upper-boundary estimate, Keller et al. (1998) estimated that between 60 and 160 public-supply wells may have been contaminated with MTBE in the State of California.

A database of all MTBE samples collected from public supplies beginning October 1989 through the present was downloaded from the California Department of Health Services publications (2002), and analyzed. Data was truncated to September 2001 to represent a similar timeframe as the other states information. Data stored in the database is a combination of ground- and surface-water samples as well as results of analysis of MTBE in samples of raw and treated waters. A total of 50,748 samples were analyzed, with 535 MTBE detections. The lowest concentration detected was 0.15 μg/L with a maximum MTBE concentration of 610 μg/L.

Connecticut

The State of Connecticut Department of Public Health is required by law to provide an annual report on results of organic chemical testing by public water systems to the Governor of the state. One of the chemicals monitored is MTBE. Only the most recent report is summarized herein (State of Connecticut Department of Public Health 2000).

In 1999, a total of 139 public water systems in 77 different towns contained organic chemicals. MTBE was detected in a total of 57 sources of public water supply in 40 of the 77 towns. Some detections within the same town were in more than one public water system. When detected, MTBE concentrations ranged from 0.7 to 110 μg/L with a median concentration of 2.7 μg/L.

Florida

The Florida Drinking Water Program has been sampling for MTBE as an unregulated contaminant since the early 1990's. The data have been stored in a database and are available on the referenced web site (Florida Department of Environmental Protection 2001). This data set indicated that a total of 8,739 samples were analyzed from 1,692 public water supplies.

MTBE was detected in 4.9 percent of all samples (428 of 8,739) and 1.2 percent (20 of 1,692) of PWS systems tested. The majority (379) of MTBE detections occurred in two public water systems. The minimum, median, and maximum detected concentrations were 0.1, 1.4, and 166 μg/L, respectively.

Illinois

CWSs in Illinois routinely sample for VOCs under the Safe Drinking Water Act monitoring program. Under Illinois' CWS Laboratory Fee Program, analyses for MTBE have been reported as part of standard laboratory methods since 1994. Approximately 80 percent of the 1,200 CWSs that supply water participate in the program, and most (1,100 of 1,200) utilize ground water as the source of drinking water.

The results reported by the Illinois Environmental Protection Agency (2000) indicated that 26 active systems have had detections of MTBE. In addition, three CWSs have had to discontinue use of wells as a result of MTBE contamination. These CWSs were located in Kankakee County, Island Lake, and East Alton, Illinois. Although concentration data were not given, the Illinois Environmental Protection Agency (2000) did indicate that most of the concentrations were unlikely to cause adverse human health effects.

Iowa

In response to growing national concern over MTBE contamination in ground water and drinking-water supplies, the Iowa Comprehensive Petroleum Underground Storage Tank Fund and the Iowa Department of Natural Resources contracted with the University of Iowa's Hygienic Laboratory to conduct analyses of water samples from PWS wells located in Iowa's vulnerable bedrock regions. As part of the Risk-Based Corrective Action process, any municipal water-supply well located within a 1-mile radius of a LUST site that is in a vulnerable bedrock region must be sampled for chemicals of concern. MTBE was added to the list of analytes to be sampled on July 1, 1999.

A total of 530 samples were collected prior to any treatment from 235 PWS systems during the second and third quarters of 1999 and analyzed (Iowa Department of Natural Resources, 2000). Similarly, 518 samples were collected and analyzed during the fourth quarter of 1999. MTBE was not detected in any sample above the quantitation limit of 15 μg/L. However, MTBE was detected below the quantitation limit in eight samples.

MTBE also has been detected in Iowa CWSs, however, not as part of the bedrock project. Cities in which MTBE has been detected included Ida Grove, Galva, and Alvord. These cities are located in northwestern Iowa. MTBE has been detected in Ida Grove's drinking-water supply each quarter since 1997. The highest concentration reported was 12 μg/L in 1998 after treatment but before blending with other source wells. MTBE was detected at a concentration of 18 μg/L in one sample from Galva's drinking-water supply in 1996, and concentrations as high as 63 μg/L in Alvord's supply were detected in 1994. Galva and Alvord have since abandoned their water-supply wells and are using a different source (Iowa Department of Natural Resources, 2000).

Kansas

News reports about MTBE in water supplies caused concern for residents of Kansas. This prompted the Kansas Department of Health and Environment (KDHE), which has routinely monitored MTBE in PWSs since 1996, to respond to these concerns (Kansas Department of Health and Environment, 2000).

From 1996 to January 2000, 27,935 water samples from 1,122 PWS wells were analyzed. MTBE was detected in 101 samples in 18 PWS wells. This equates to MTBE being detected in 1.6 percent of the 1,122 PWS wells. Concentrations ranged from 0.5 to 1,250 μg/L (Greg Hattan, Kansas Department of Health and Environment, written commun., 2002). In response to these findings, KDHE has installed three treatment systems at PWS wells where MTBE was detected at higher concentrations and plans to install a fourth treatment system.

Maine

In response to public awareness and concern for the presence of MTBE in Maine's water resources, Maine Governor Angus King directed state health and environmental agencies to undertake an assessment of the occurrence of MTBE in Maine's drinking-water supplies by sampling all (830) public and 1,000 private household water supplies. When completed, the assessment sampled 793 of the 830 regulated nontransient PWSs and 951 private household water supplies (Maine Bureau of Health, 1998).

MTBE was detected in 125 (16 percent) of the 793 PWSs that were sampled. All detected concentrations were less than Maine's MCL of 35 μg/L. About 6 percent of the samples had detected concentrations between 1 and 35 μg/L, and in 93.9 percent of the samples, MTBE was either not detected or less than 1 μg/L. Public water supplies serving businesses or mobile home parks were found to be about twice as likely to have detectable levels of MTBE as compared with community water supplies and schools. Location of the water supply in areas with required RFG use and high population density were both associated with detectable levels of MTBE. Factors that were found not to be associated with MTBE detection included: (1) type of well or water supply, and (2) proximity to gasoline storage tanks.

Population density was a significant risk factor for PWSs within areas where RFG use was required. In areas of high population density (greater than 180 people per square mile), the risk of MTBE detection was 4.1 times higher in areas where RFG use was required compared to areas where it was not. In areas of low population density, the risk of MTBE detection was 1.7 times higher in areas where RFG use was required compared to areas where it was not.

MTBE was detected in 150 (15.8 percent) of the 951 private wells sampled. Ten (1.1 percent) of the 951 sampled wells contained MTBE at concentrations greater than the Maine MCL of 35 μg/L. Extrapolation of these findings indicates that an estimated 1,400 to 5,200 private wells may contain MTBE at concentrations greater than or equal to 35 μg/L. Sixty-three (6.6 percent) of the private wells contained MTBE at concentrations between 1 and 35 μg/L. Thus, about 92 percent of the wells did not contain detectable levels of MTBE or had concentrations less than 1 μg/L.

Population density was a significant risk factor for private household water supplies within areas where RFG use was required. In areas of high population, the risk of MTBE detection in private household water supplies was 1.3 times higher in areas where RFG use is required compared to areas where it was not. In areas of low population density, the risk of MTBE detection was 2.0 times higher in areas where RFG use was required compared to areas where it was not.

Maryland

The Maryland Department of the Environment (2001) has been periodically monitoring public water systems, specifically community and nontransient noncommunity public water systems, for MTBE since 1995. MTBE was detected in 85 public water systems (7.8 percent) out of 1,084 sampled. Eleven of these systems contained MTBE concentrations greater than 20 μg/L, and 10 of the 85 systems now have alternative sources or the concentrations have since declined to levels below 20 μg/L.

In addition, all LUSTs that affect ground water within the state have water-quality monitoring. At the time these data were obtained from the Maryland Department of Environment web site (June 8, 2001), 270 domestic wells had been affected by MTBE.

Michigan

In March 2000, the Michigan Department of Environmental Quality's (MDEQ) Storage Tank Division released a fact sheet summarizing MTBE specific information (Michigan Department of Environmental Quality, 2000). In 1996, Michigan began enforcing a low-vapor-pressure requirement in summertime fuel to control ozone levels for gasoline sold between June 1 and September 15. Rather than supplying RFG containing MTBE or ethanol, the refiners supplying gasoline during this timeframe opted to meet the low-vapor-pressure requirement by removing some of the more volatile compounds from the gasoline. The chemical composition of gasoline sold throughout the year is monitored by the Michigan Department of Agriculture. Gasoline samples are randomly collected and analyzed; however, the total number of samples collected each year is not known.

The results of the survey completed during 1998 indicated that 8 and 5 percent of the fuels sampled contained ethanol and MTBE, respectively. Concentrations of ethanol ranged from 9 to 10 percent by volume and a "very small fraction" contained MTBE in concentrations greater than 12 percent by volume. Low concentrations (less than 2.2 percent by volume) of MTBE were thought to be due to MTBE in the pipeline distribution system mixing with the next fuel to be transported. MTBE in gasoline at levels greater than 2.2 percent by volume was thought to be present as an octane enhancer.

Although MTBE was present in only about 5 percent of gasoline sold in Michigan, the MDEQ identified MTBE at several LUST sites. The Drinking Water and Radiological Protection Division (DWRPD) also detected MTBE in drinking water. The DWRPD has been analyzing samples for MTBE since 1987. From October 1, 1987, to September 30, 1999, 31,557 water samples from 18,046 community, non-community, and private water wells were analyzed by the DWRPD. Results indicated that MTBE was detected in 903 samples from 542 locations. This equates to 2.9 percent of all samples and 3.0 percent of the community, non-community, and private wells sampled. The reporting level used for these analyses was not presented but was assumed to be 1.0 μg/L for most analyses. For those samples in which MTBE was detected, 3.2 percent (29 of 903) contained MTBE at concentrations greater than 240 μg/L, 9.1 percent (82 of 903) contained MTBE at concentrations between 40 and 240 μg/L, and 87.7 percent (792 of 903) contained MTBE at concentrations between 1 and 40 μg/L (Michigan Department of Environmental Quality, 2000).

Missouri

In response to a request from State Senator David Klarich, the Missouri Department of Natural Resources (MDNR) reported their perspective on a State resolution that requires them to assess the extent of possible environmental contamination of MTBE (Missouri Department of Natural Resources, 2000). The MDNR has assessed MTBE in the environment since 1992 and has analyzed for MTBE in public drinking-water sources since 1994. There are approximately 1,685 public water supplies and non-transient community systems in Missouri, each of which are monitored at least once every 3 years for ground-water supplies or annually for surface-water supplies. Some water suppliers are required to analyze samples more frequently, such as in the St. Louis, Missouri, non-attainment area, whereas the larger systems do so voluntarily.

As of February 7, 2000, MTBE had been detected in two public water supplies. The MDNR indicated that the contamination was from underground tanks leaking premium-grade gasoline rather than RFG.

New Jersey

The continued interest in MTBE in the environment prompted the New Jersey Department of Environmental Protection (1999) to summarize and evaluate, in part, information on the use and occurrence of MTBE in New Jersey. Some of the information presented specifically related to the occurrence of MTBE in public and domestic water supplies.

The Bureau of Safe Drinking Water has been collecting data on MTBE in public water supplies since 1997. Results of samples collected from about 400 CWSs from July 1997 to September 1998 were presented. At a reporting level of 0.5 μg/L, results indicated that MTBE was detected in 59 (15 percent) of the CWSs with a maximum concentration of 8.9 μg/L. The samples analyzed typically were finished drinking water.

The majority of information on MTBE in domestic wells was obtained in response to a potential or existing contamination problem associated with the well. As such, results from these wells would be biased high making it difficult to define the overall presence of MTBE in these types of wells. However, one assessment of domestic wells was completed in which the wells sampled were selected on a random basis. The wells were selected and ultimately sampled from four main areas of New Jersey: the New England physiographic province (Highlands); the Piedmont; the Kirkwood-Cohansey aquifer in southern New Jersey; and the crystalline rock aquifer surrounding Cranberry Lake in Sussex County. MTBE was detected in all four areas (reporting level of 0.1 μg/L). For all four areas combined, MTBE was detected in 35.6 percent of the domestic wells (37 of 104 wells sampled). Concentrations typically were low, with minimum and maximum detected concentrations of 0.1 and 30.2 μg/L, respectively.

Wisconsin

The State of Wisconsin has systematically analyzed for MTBE in ground water used as a source of drinking water since 1990. All results are entered into the Wisconsin Department of Natural Resources (DNR) Groundwater Retrieval Network (GRN) database.

In response to the national interest in MTBE in sources of drinking water, the DNR summarized MTBE data obtained from the GRN database (Wisconsin Department of Natural Resources Bureau for Remediation and Redevelopment, 2000). Results indicate that MTBE was detected in 99 (4.4 percent) of 2,271 wells sampled. Most (96) of the MTBE detections were found in private residential wells because municipal water-supply system purveyors in Wisconsin are not required to monitor for MTBE. Two wells that were described as "municipal-community" wells and one well that was described as "community, other than municipal" also had detected concentrations of MTBE. The statistical summary of MTBE concentrations were not reported by the DNR. However, the maximum concentration reported (1,700 μg/L) was from a private-residential well.

Literature Survey Summary

Previously (1997), an exhaustive review of MTBE in water was completed as part of an Interagency Assessment of Oxygenated Fuels. Results of the Interagency literature review concluded that it was not possible to characterize MTBE in sources of drinking water due to limited data available at that time. As such, reviewed literature for this report focused on those assessments completed subsequent to the Interagency review.

Specifically, the review for this report focused on public and private water-supply studies that are national, regional, or statewide in scope. In some cases, results are reported from summaries of information provided by states that were based on the state's normal compliance monitoring rather than a designated assessment. Overall, three national, two regional, and 13 statewide assessments were reviewed.

A summary of the occurrence, frequency, and concentration of MTBE detected in drinking water is presented in Table 2.1. It is important to note that it was usually not possible to compare findings between assessments, in part, due to different study objectives, different reporting levels, and different types of water sampled including untreated drinking water, finished drinking water, and ambient water. In addition, many samples reported in this literature review were collected in response to consumer complaints. This potentially results in detected concentrations that may be high in comparison to ambient levels. The inconsistencies among assessments studies further support the Interagency conclusion that an overall characterization of MTBE in drinking water was not possible on the basis of existing information and that further investigations, such as the AwwaRF-sponsored assessment, were necessary.

Overall, results of this literature review indicate that MTBE has been detected in public and (or) private drinking-water supplies in 36 states, which are listed in Table 2.2. Detected concentrations of MTBE ranged from 0.1 to 23,000 µg/L. In some cases, maximum concentrations were reported only as greater than a certain concentration. Thus, it is not known what those maximum concentrations were. Many times, it was not possible to determine median detected concentrations because only minimum and maximum concentrations were reported. However, when median detected concentrations were available, they were typically low--less than 5.0 µg/L. Similarly, the number of samples collected or the number of systems from which samples were collected was not always known. Thus, it was not always possible to determine or calculate detection frequencies. Regardless, these assessments did indicate that: (1) MTBE occurred in public drinking-water systems supplied by both ground and surface water, and concentrations generally were less than 20 µg/L; (2) population density and RFG use were significant risk factors for MTBE detection in water supplies; (3) type of well and proximity to gasoline storage tanks did not seem to be associated with MTBE detection; and (4) more data are needed before the extent of MTBE in drinking water and its resulting importance as a route of human exposure can be determined.

Table 2.1

Summary of methyl *tert*-butyl ether (MTBE) in sources of drinking water reported in national, regional, and statewide assessments

Reference	Type of water sampled	Number of samples collected	Reporting level (μg/L)	Number of MTBE detections	Detection frequency (percent)	Detected concentration (μg/L)		
						Minimum	Median	Maximum
National Assessments								
Zogorski and others, 1997	Ground water and drinking water obtained from ground water[1]	1,516	0.2 – 1.0	76	5.0	0.2	0.6	[2]23,000
	Public ground-water supplies[1,3]	NK	0.2 – 1.0	2	NK	0.3	1.1	1.8
	Private ground-water supplies[1,3]	NK	0.2 – 1.0	11	NK	0.2	.5	2.2
	Public surface-water supplies[4]	NK	NK	1	NK	1.0	1.0	1.0
	Public ground-water supplies[4]	NK	NK	241	NK	0.7	3.0	17,800
	Private ground-water supplies[4]	NK	NK	680	NK	1.0	27	17,000
M.J. Moran, U.S.Geological Survey, 2002a	Private ground-water supplies	1,335	0.2	30	2.2	0.2	.7	30.2
Regional Assessments								
Gullick and LaChevallier, 2000	Public ground-water supplies	[5]1,349	0.5	30	[6]8.8	0.5	NK	14.1
	Public surface-water supplies	[7]200	0.5	8	[8]8.7	0.7	NK	25.1
Grady and Casey, 2001	Finished drinking water	[9]5,510	0.5 – 10	106	[10]8.9	0.26	NK	210
Statewide Assessments								
Alabama Department of Environmental Management, 2001[11]	Public surface-water supplies	[12]87	.5 – 2.0	0	0	ND	ND	ND
	Public ground-water supplies	966	0.5 – 2.0	5	0.5	0.74	NK	8.39
Keller and others, 1998 (California)	Public surface-water supplies	[13]3,000	NK	NK	[14]46.7	NK	NK	[15]>14
	Public ground-water supplies	[16]2,988	NK	35	1.2	NK	NK	NK
California Department of Health Services, 2002	Public water supplies	50,748	NK	535	1.1	0.15	3.6	610

(continued)

Table 2.1 (continued)

Reference	Type of water sampled	Number of samples collected	Reporting level (µg/L)	Number of MTBE detections	Detection frequency (percent)	Detected concentration (µg/L)		
						Minimum	Median	Maximum
State of Connecticut Department of Public Health, 2000	Public water supplies	NK	0.5 – 2.0	57	NK	0.7	2.7	110
Florida Department of Environmental Protection 2001	Public water supplies	[17]8,739	NK	428	[18]1.2	0.1	1.4	166
Illinois Environmental Protection Agency, 2000	Public water supplies	[19]960	0.5 – 1.0	26	[18]2.7	NK	NK	NK
Iowa Department of Natural Resources, 2000	Public ground-water supplies	[20]1,048	15	NK	[21]NK	0.46	NK	63
Kansas Department of Health and Environment, 2000	Public water supplies	[22]27,935	NK	101	[18]1.6	0.5	NK	1,250
Maine Bureau of Health, 1998	Public water supplies	[23]793	0.1	125	15.8	NK	NK	[24]<35
	Private ground-water supplies	[23]951	0.1	150	15.8	NK	NK	[25]>35
Maryland Department of the Environment, 2001	Public water supplies	1,084	0.5	[26]85	[27]7.8	0.5	NK	[28]>20
Michigan Department of Environmental Quality, 2000	Public and private water supplies	[29]31,557	1.0	903	[6]3.0	1.0	NK	[30]>240
Missouri Department of Natural Resources, 2000	Public water supplies	[31]2,700	5	2	[6]7	NK	NK	NK
New Jersey Department of Environmental Protection 1999	Public water supplies	[32]400	0.5	59	14.8	NK	NK	8.4
	Private ground-water supplies	104	0.1	37	35.6	0.10	[33]0.48	30.2
Wisconsin Department of Natural Resources Bureau for Remediation and Redevelopment, 2000	Private ground-water supplies	2,271	12	[34]99	4.4	NK	NK	1,700

[µg/L, micrograms per liter; NK, not known; ND, not detected; >, greater than]

1 Data provided by the U.S. Geological Survey for the referenced study.

2 This concentration was detected in ambient ground water, not in a drinking-water supplies.

3 These data are a subset of the ground water and drinking water obtained from the U.S. Geological Survey ground-water category listed above.

4 Data requested by and provided to the U.S. Environmental Protection Agency for the referenced study.

5 1,349 samples were collected from 342 wells; 270 of the 342 wells were sampled before treatment, and 72 were sampled as treated water.

6 Represents the number of wells with at least one detected concentration.

7 200 samples were collected from 92 surface-water sites.

8 Represents the percent of intakes with at least one detected concentration.

9 5,510 samples were collected from 1,194 community water-supply systems.

10 Detection frequency represents the number of CWSs with at least one detected concentration using reporting levels ranging from 0.5 to 10μg/L.

11 Efforts were made to sample all public drinking-water supplies in Alabama (701).

12 Represents the number of surface-water sources sampled. Most were sampled quarterly, however, the number of samples collected is not known.

13 About 3,000 samples (exact number is not known) were collected from a total of 105 surface-water sources. The majority of samples were collected at only 25 of the 105 sources.

14 Represents the number of sources with at least one detection.

15 The maximum detected concentration was not presented. However, the report stated that 13 surface-water sources had concentrations greater than 14μg/L.

16 2,988 public-supply wells were sampled. It is unknown if any wells were sampled multiple times.

17 8,739 water samples were collected from 1,692 public water systems.

18 Represents the number of public water-supply systems with at least one detection of MTBE.

19 960 is estimated based on 80 percent of 1,200 community water suppliers reported to participate in the study. Most (1,100 of 1,200) community water suppliers utilize ground water as the primary source of drinking water.

20 1,048 samples were collected from 235 public water-supply systems. All wells were located within 1 mile of a LUST site.

21 MTBE was detected in eight samples below the minimum reporting level (MRL).

22 27,935 water samples were collected from 1,122 public water supplies. MTBE was detected in 18 of the public water supplies.

23 Individual wells were sampled once.

24 All detected concentrations were less than 35μg/L, however, the maximum concentration was not presented.

25 1.1 percent of the sampled wells contained concentrations that were greater than 35μg/L, however, the actual concentrations was not presented.

26 MTBE was detected in 85 systems. It is not known how many total samples were collected or how many samples contained MTBE.

27 According to the data on the web site, a total of 1,084 systems were sampled.

28 11 systems were found to contain MTBE at concentrations greater than 20μg/L. The maximum concentration detected was not presented.

29 31,557 water samples were collected from 18,046 community, non-community, and private wells.

30 The maximum detected concentration of MTBE is not known, however, it was presented to be greater than 240μg/L.

31 Approximately 2,700 samples were collected from about 20 public water supplies. Two were found to contain MTBE.

32 Actual number of samples is unknown; samples were submitted from approximately 400 CWSs.

33 Represents the average median concentration from four sampling areas.

34 Actual number of samples with MTBE detected is unknown. This represents the number of wells rather than number of samples.

Table 2.2
States with detected concentrations of methyl *tert*-butyl ether in public and (or) private drinking-water supplies

Alabama	Kansas	North Carolina
Arkansas	Maine	Oregon
California	Maryland	Pennsylvania
Colorado	Massachusetts	Rhode Island
Connecticut	Michigan	South Carolina
Delaware	Minnesota	Tennessee
Florida	Missouri	Texas
Georgia	Montana	Washington
Idaho	New Hampshire	West Virginia
Illinois	New Jersey	Wisconsin
Indiana	New Mexico	Vermont
Iowa	New York	Virginia

A REVIEW OF LITERATURE REGARDING THE TASTE AND ODOR PROPERTIES OF MTBE IN DRINKING WATER AND TREATMENT TECHNOLOGIES

Purpose and Scope

The potential for MTBE to intrude into ground and surface water sources presents a concern that is secondary to health effects, but to which consumers are acutely aware - the aesthetic quality of the water. MTBE has a distinctive and unpleasant odor that is often compared to turpentine. At dilute concentrations, descriptors such as "solvent" and "sweet solvent" have been used to characterize water tainted with MTBE. Several studies have been conducted to define the taste and odor (T&O) thresholds for MTBE. A definitive threshold has been somewhat elusive, as the studies vary in design, the number of participants, and intrinsically are very subjective in nature. This review provides a summary of the work that has been completed to date.

MTBE Taste and Odor Threshold Studies

Six studies on the aesthetic implications of MTBE in drinking water are known: 1) The Odor and Taste Threshold Studies Performed with Methyl *Tertiary*-Butyl Ether and Ethyl *Tertiary*-Butyl Ether (ARCO Chemical 1993); 2) Odor Threshold Studies Performed with Gasoline and Gasoline Combined with MTBE, ETBE, and TAME (American Petroleum Institute (API) 1994); 3) Taste and Odour Threshold Concentrations of Potential Potable Water Contaminants (Young et al. 1995); 4) Threshold Odor Concentrations of MTBE and other Fuel Oxygenates (Shen et al. 1997); 5) MTBE: Taste and Odor Threshold Determinations Using the Flavor Profile Method (Dale et al. 1997); and 6) Implications of an MTBE Odor Study for Setting Drinking Water Standards (Stocking et al. 2001).

The first five studies reported taste and odor "thresholds" for MTBE ranging from 13 to 140 μg/L. Each of the five studies were conducted under differing conditions; however, the participants were all members of the scientific community, and many were trained specialists in aroma and flavor analysis. In addition, the number of panel participants was small. Some of the authors concluded that a larger panel of consumer participants would provide a better representation of the general public's perception of MTBE in water. Due to the concern over health effects, no further taste studies have been conducted. The latest study gathered a larger participant base to assess the odor threshold of MTBE in water (Stocking et al. 2001).

The ARCO study (1993) involved a six-member panel. The panelists evaluated the taste and odor of water spiked at over 106 μg/L. A linear regression was applied to the results to obtain threshold levels of 149 μg/L (taste) and 95 μg/L (odor). More recent information has revealed that threshold testing for MTBE should be conducted at levels well below 100 μg/L.

API commissioned a second study (1994), conducted by the same group that did the testing for ARCO (1993). Using a similar test design, seven panel members evaluated the taste and odor of MTBE in water at concentrations between 23 μg/L and 740 μg/L. Again, a linear regression was applied which yielded threshold values of 39-44 μg/L (taste) and 41-46 μg/L (odor).

Young et al. (1995) published findings pertaining to the taste and odor properties of a wide variety of chemical compounds, including MTBE. The results were based on the geometric mean of the individual results from five to seven panelists. The thresholds were determined to be 48 μg/L (taste) and 34 μg/L (odor).

The Orange County Water District of Southern California conducted a study to determine odor thresholds of MTBE and other oxygenates in drinking water (Shen et al. 1997). The MTBE odor studies were conducted in the concentration range of 0 to 159 μg/L, and at temperatures of 25° C, 40° C, and 60° C. The geometric means were calculated to derive odor thresholds ranging from 13.5 to 45.4 μg/L.

Another study was conducted at the MWDSC to study taste and odor thresholds of MTBE in drinking water (Dale et al. 1997). The threshold studies utilized nine trained flavor profile panelists, using the triangle test and a statistical determination of the concentrations at which there occurred a 60 percent probability of detection of MTBE versus odor-free water. This test produced threshold ranges of 24 to 37 μg/L (taste), and 43 to 71 μg/L (odor).

The study by Stocking et al. (2001) convened a panel of 57 untrained consumers. The study used an odor testing protocol based on the American Society for Testing and Materials method E679-91 (ASTM 1991). The odor threshold for MTBE was determined by this study to be 15 μg/L. That value has become accepted by many water utilities to serve as a guideline to assess the aesthetic acceptability of drinking water. Though the US Environmental Protection Agency has not promulgated a secondary action level for MTBE, many states have adopted either an action level or a secondary maximum contaminant level for MTBE (both of which are unenforceable standards).

MTBE Treatment Technologies

Conventional treatment techniques (e.g., coagulation, flocculation, sedimentation and filtration) are not effective in removing MTBE from drinking water. Alternative existing and emerging technologies are costly and difficult to optimize. The California MTBE Research Partnership has conducted studies to evaluate the following treatment technologies for the removal of MTBE from drinking water: air stripping, advanced oxidation processes, granular activated carbon, and synthetic resin sorbents (National Water Research Institute 1999). Some cities are using air stripping in combination with another treatment (e.g., granular activated carbon) to remove MTBE.

CHAPTER 3
METHODS

STUDY DESIGN

Purpose and Scope

This section describes the design of a national survey of MTBE in public-supply drinking-water sources. The survey consisted of two independent stages designed to provide representative sampling of all CWSs in the United States (Random Survey) and to improve understanding of the temporal variability of MTBE in selected water sources (Focused Survey). The Random Survey was designed to provide representative information on the frequency of detection, concentration, and distribution of MTBE, other gasoline oxygenates, and other VOCs in 1,000 untreated, ground- and surface-water sources of drinking water used by CWSs in the United States. The Focused Survey was designed to provide an improved understanding of the frequency of detection, concentration, and temporal variability of MTBE, other gasoline oxygenates, their degradation products, and other VOCs in about 140 sources of untreated water with suspected or known MTBE contamination. The Focused Survey design considered, in part, factors that appear to be related to frequent detection of MTBE and other VOCs, when selecting CWSs for sampling. Samples were collected to better characterize the variability in the detection and concentrations of a wide range of VOCs and oxygenate by-products.

This report also describes the logic and statistical analysis used to design a stratified selection scheme to minimize bias of CWSs for the Random Survey. Similarly, justification for selection of CWSs for the Focused Survey also is provided.

Overview

Two different but complimentary survey designs were developed. For the Random Survey, statistical analysis was required to ensure an unbiased distribution of CWSs by state, source type (ground water or surface water), and population served. The Focused Survey, which used a fixed-interval sampling strategy, was conducted in two parts. Samples were collected to characterize possible temporal VOC variations by sampling wells twice in the year, reservoirs and lakes quarterly, and rivers eight times in the year. In the first part of the Focused Survey, CWSs in metropolitan areas that serve large populations (greater than 50,000) in RFG (high MTBE-use) areas where MTBE was suspected of contaminating drinking-water sources were solicited for participation. Other CWS sources suspected of having concentrations of MTBE were selected by using the following criteria: sources being near leaking underground storage tanks (LUSTs), in or near former RFG areas, and high watercraft use on surface-water sources. Second, the Focused Survey built upon data from the results of the Random Survey, as CWSs with known concentrations of MTBE were selected for inclusion in the Focused Survey. Selection of additional CWSs with known concentrations of MTBE for inclusion in the Focused Survey were made through various sources including state environmental and drinking-water reports, discussions with state officials, and other sources.

Ground-water field equipment blanks were not collected because permanently installed pumps were used to collect the samples. Surface-water equipment blanks were collected at approximately 70 percent of the sites. Field duplicate samples were collected when VOCs were known to occur on the basis of previous sampling at the sites, and were distributed evenly among the ground-water and surface-water sites. Physical and chemical properties, including specific conductance, pH, air and water temperature, and dissolved oxygen, were measured on-site prior to sampling ground water to ensure that representative aquifer water was being sampled. Similarly, these properties were measured prior to sampling surface water to document local sampling conditions.

Random Survey

The Random Survey was designed to determine the frequency of detection and the range in concentrations of MTBE, three other ether gasoline oxygenates, and 62 other VOCs in drinking-water sources through a representative random sampling of 954 CWSs. This was determined using samples collected from a statistically stratified design. Table 3.1 shows the 66 compounds, organized by compound type and lists the compound name, CAS number, IUPAC name, common name, and chemical formula. Table 3.2 shows the compounds with the method detection limit (MDL), minimum reporting level (MRL) and regulatory information (MCLs, health advisories [HAs], and drinking water CCL).

Table 3.1
Analytes- compound names

Compound Name (used throughout report)	CAS #	IUPAC NAME	Other Common names	Chemical Formula
Ether Oxygenates				
tert-Amyl methyl ether (TAME)	994-05-8	2-Methoxy-2-methylbutane		$C_6H_{14}O$
Ethyl *tertiary* butyl ether (ETBE)	637-92-3	2-Ethoxy-2-methylpropane		$C_6H_{14}O$
Methyl *tert*-butyl ether (MTBE)	1634-04-4	2-Methoxy-2-methylpropane		$C_5H_{12}O$
Diisopropyl ether (DIPE)	108-20-3	2,2'-oxybis[propane]		$C_6H_{14}O$
Gasoline Compounds		Gasoline Compounds		
Benzene	71-43-2	Benzene		C_6H_6
sec-Butylbenzene	135-98-8	(1-Methylpropyl)benzene		$C_{10}H_{14}$
tert-Butylbenzene	98-06-6	(1,1-Dimethylethyl)benzene		$C_{10}H_{14}$
n-Butylbenzene	104-51-8	n-Butylbenzene	1-phenylbutane	$C_{10}H_{14}$
Ethylbenzene	100-41-4	Ethylbenzene	phenylethane	C_8H_{10}
Naphthalene	91-20-3	Naphthalene		$C_{10}H_8$
Toluene	108-88-3	Methylbenzene		C_7H_8
1,3,5-Trimethylbenzene	108-67-8	1,3,5-Trimethylbenzene	mesitylene	C_9H_{12}
m,p-Xylene	108-38-3	1,3-Dimethylbenzene & 1,4-Dimethylbenzene	m-xylene & p-xylene	C_8H_10
o-Xylene	106-42-3/ 95-47-6	1,2-Dimethylbenzene		C_8H_{10}
Solvents/Organic Synthesis				
Acrylonitrile	107-13-1	2-Propenenitrile		C_3H_3N
Bromobenzene	108-86-1	Bromobenzene	phenyl bromide	C_6H_5Br
Bromochloromethane	74-97-5	Bromochloromethane	methylene chlorobromide	CH_2BrCl
Carbon Tetrachloride	56-23-5	Tetrachloromethane		CCl_4

(continued)

Table 3.1 (continued)

Compound Name (used throughout report)	CAS #	IUPAC NAME	Other Common names	Chemical Formula
Chlorobenzene	108-90-7	Chlorobenzene	monochlorobenzene	C_6H_5Cl
Chloroethane	75-00-3	Chloroethane	ethyl chloride	C_2H_5Cl
2-Chlorotoluene	95-49-8	1-Chloro-2-methylbenzene	o-chlorotoluene	C_7H_7Cl
4-Chlorotoluene	106-43-4	1-Chloro-4-methylbenzene	p-chlorotoluene	C_7H_7Cl
Dibromomethane	74-95-3	Dibromomethane	methylene bromide	CH_2Br_2
1,2-Dichlorobenzene	95-50-1	1,2-Dichlorobenzene	o-dichlorobenzene	$C_6H_4Cl_2$
1,3-Dichlorobenzene	541-73-1	1,3-Dichlorobenzene	m-dichlorobenzene	$C_6H_4Cl_2$
1,1-Dichloroethane	75-34-3	1,1-Dichloroethane	ethylidene chloride	$C_2H_4Cl_2$
1,2-Dichloroethane	107-06-2	1,2-Dichloroethane	ethylene dichloride	$C_2H_4Cl_2$
1,1-Dichloroethene	75-35-4	1,1-Dichloroethene	vinylidene chloride	$C_2H_2Cl_2$
cis-1,2-Dichloroethene	156-59-2	cis-1,2-Dichloroethene	(Z)-1,2-dichloroethene	$C_2H_2Cl_2$
trans-1,2-Dichloroethene	156-60-5	trans-1,2-Dichloroethene	(E)-1,2-dichloroethene	$C_2H_2Cl_2$
1,2-Dichloropropane	78-87-5	1,2-Dichloropropane	propylene dichloride	$C_3H_6Cl_2$
1,3-Dichloropropane	142-28-9	1,3-Dichloropropane	trimethylene dichloride	$C_3H_6Cl_2$
2,2-Dichloropropane	594-20-7	2,2-Dichloropropane		$C_3H_6Cl_2$
1,1-Dichloropropene	563-58-6	1,1-Dichloropropene		$C_3H_4Cl_2$
Hexachlorobutadiene	87-68-3	1,1,2,3,4,4-Hexachloro-1,3-butadiene		C_4Cl_6
Hexachlororethane	67-72-1	1,1,1,2,2,2-Hexachloroethane	carbon hexachloride	C_2Cl_6
p-Isopropyltoluene	99-87-6	1-Isopropyl-4-methylbenzene		$C_{10}H_{14}$
Isopropylbenzene	98-82-8	1-Methylethyl)benzene		C_9H_{12}
Methylene Chloride	75-09-2	Dichloromethane		CH_2Cl_2
Methyl ethyl ketone (MEK)	78-93-3	2-Butanone		C_4H_8O
n-Propylbenzene	103-65-1	n-Propylbenzene	1-phenylpropane	C_9H_{12}
Styrene	100-42-5	Ethenylbenzene		C_8H_8
1,1,1,2-Tetrachloroethane	630-20-6	1,1,1,2-Tetrachloroethane		$C_2H_2Cl_4$

(continued)

Table 3.1 (continued)

Compound Name (used throughout report)	CAS #	IUPAC NAME	Other Common names	Chemical Formula
1,1,2,2-Tetrachloroethane	79-34-5	1,1,2,2-Tetrachloroethane		$C_2H_2Cl_4$
Tetrachloroethene	127-18-4	Tetrachloroethene	perchloroethene, PCE	C_2Cl_4
1,2,3-Trichlorobenzene	87-61-6	1,2,3-Trichlorobenzene		$C_6H_3Cl_3$
1,2,4-Trichlorobenzene	120-82-1	1,2,4-Trichlorobenzene		$C_6H_3Cl_3$
1,1,1-Trichloroethane	71-55-6	1,1,1-Trichloroethane	methylchloroform	$C_2H_3Cl_3$
1,1,2-Trichloroethane	79-00-5	1,1,2-Trichloroethane	vinyl trichloride	$C_2H_3Cl_3$
Trichloroethene (TCE)	79-01-6	Trichloroethene	trichloroethylene	C_2HCl_3
1,2,3-Trichloropropane	96-18-4	1,2,3-Trichloropropane	allyl trichloride	$C_3H_5Cl_3$
Vinyl bromide	593-60-2	Bromoethene		C_2H_3Br
Vinyl chloride	75-01-4	Chloroethene		C_2H_3Cl
Refrigerants				
Chloromethane	74-87-3	Chloromethane	methyl chloride	CH_3Cl
Dichlorodifluoromethane	75-71-8	Dichlorodifluoromethane	CFC 12	CCl_2F_2
Trichlorofluoromethane	75-69-4	Trichlorofluoromethane	CFC 11	CCl_3F
1,1,2-Trichloro-1,2,2-trifluoroethane	76-13-1	1,1,2-Trichloro-1,2,2-trifluoroethane	CFC 113	$C_2Cl_3F_3$
Fumigants				
Bromomethane	74-83-9	Bromomethane	methyl bromide	CH_3Br
1,4-Dichlorobenzene	106-46-7	1,4-Dichlorobenzene	p-dichlorobenzene	$C_6H_4Cl_2$
cis-1,3-Dichloropropene	10061-01-5	cis-1,3-Dichloropropene	(Z)-1,3-dichloropropene	$C_3H_4Cl_2$
trans-1,3-Dichloropropene	10061-02-6	trans-1,3-Dichloropropene	(E)-1,3-dichloropropene	$C_3H_4Cl_2$
Disinfection By-Product				
Bromodichloromethane	75-27-4	Bromodichloromethane	dichlorobromomethane	$CHBrCl_2$
Bromoform	75-25-2	Tribromomethane		$CHBr_3$
Chlorodibromomethane	124-48-1	Dibromochloromethane		$CHBr_2Cl$
Chloroform	67-66-3	Trichloromethane		$CHCl_3$

Table 3.2
Volatile organic compound analytes

Analyte	MDL (μg/L)	MRL (μg/L)	USEPA MCL (μg/L)	HA (μg/L)	USEPA DWCCL[†]
Ether Oxygenates					
tert-Amyl methyl ether	0.025	0.2			
Ethyl *tertiary*-butyl ether	0.034	0.2			
Methyl *tertiary*-butyl ether	0.039	0.2		20-40[*]	Yes
Diisopropyl ether	0.073	0.2			
VOCs on DWCCL[†]					
Bromobenzene	0.029	0.2			Yes
Bromomethane	0.084	0.2		10	Yes
1,1-Dichloroethane	0.036	0.2			Yes
1,3-Dichloropropane	0.029	0.2			Yes
2,2-Dichloropropane	0.056	0.2			Yes
1,1-Dichloropropene	0.060	0.2			Yes
cis-1,3-Dichloropropene	0.024	0.2			Yes
trans-1,3-Dichloropropene	0.026	0.2			Yes
Hexachlorobutadiene	0.057	0.2		1	Yes
p-Isopropyltoluene	0.037	0.2			Yes
Naphthalene	0.055	0.2		20	Yes
1,1,2,2-Tetrachloroethane	0.026	0.2			Yes
1,2,4-Trimethylbenzene	0.022	0.2			Yes
Other VOCs					
Acrylonitrile	0.098	0.2			
Benzene	0.029	0.2	5		
Bromochloromethane	0.036	0.2		90	
Bromodichloromethane	0.018	0.2	‡		
Bromoform	0.022	0.2	‡		
sec-Butylbenzene	0.044	0.2			
tert-Butylbenzene	0.037	0.2			
n-Butylbenzene	0.047	0.2			
Carbon tetrachloride	0.049	0.2	5		
Chlorobenzene	0.032	0.2	100	100	
Chloroethane	0.095	0.2			
Chloroform	0.024	0.2	‡		
Chloromethane	0.105	0.2		3	
2-Chlorotoluene	0.033	0.2		100	
4-Chlorotoluene	0.030	0.2		100	
Chlorodibromomethane	0.016	0.2	‡	60	
Dibromomethane	0.028	0.2			
1,2-Dichlorobenzene	0.037	0.2	600	600	
1,3-Dichlorobenzene	0.029	0.2		600	

(continued)

Table 3.2 (continued)

Analyte	MDL (µg/L)	MRL (µg/L)	USEPA MCL (µg/L)	HA (µg/L)	USEPA DWCCL†
1,4-Dichlorobenzene	0.033	0.2	75	75	
Dichlorodifluoromethane	0.121	0.2		1,000	
1,2-Dichloroethane	0.029	0.2	5		
1,1-Dichloroethene	0.082	0.2	7	7	
cis-1,2-Dichloroethene	0.024	0.2	70	70	
trans-1,2-Dichloroethene	0.040	0.2	100	100	
1,2-Dichloropropane	0.028	0.2	5		
Ethylbenzene	0.033	0.2	700	700	
Hexachloroethane	0.086	0.2		1	
Isopropylbenzene	0.040	0.2			
Methylene chloride	0.021	0.2	5		
Methyl ethyl ketone	0.645	2.0			
n-Propylbenzene	0.043	0.2			
Styrene	0.026	0.2	100	100	
1,1,1,2-Tetrachloroethane	0.02	0.2		70	
Tetrachloroethene	0.049	0.2	5		
Toluene	0.025	0.2	1,000	1,000	
1,2,3-Trichlorobenzene	0.042	0.2			
1,2,4-Trichlorobenzene	0.042	0.2	70	70	
1,1,1-Trichloroethane	0.045	0.2	200	200	
1,1,2-Trichloroethane	0.014	0.2	5	3	
Trichloroethene	0.034	0.2	5		
Trichlorofluoromethane	0.096	0.2		2000	
1,2,3-Trichloropropane	0.027	0.2		40	
1,1,2-Trichloro-1,2,2-	0.099	0.2			
1,3,5-Trimethylbenzene	0.026	0.2			
Vinyl bromide	0.084	0.2			
Vinyl chloride	0.082	0.2	2		
m,p-Xylene	0.065	0.2	10,000	10,000	
o-Xylene	0.028	0.2	10,000	10,000	

MDL = method detection limit; MRL = minimum reporting level, MCL = maximum contaminant level, HA = health advisory

* USEPA Drinking Water Advisory based on taste and odor (U. S. Environmental Protection Agency 2000b)

† DWCCL = USEPA Drinking-Water Candidate Contaminant List (U. S. Environmental Protection Agency 1998a)

‡ total for trihalomethanes cannot exceed 80 µg/L

The results of the VOC analyses have a binomial distribution because a compound either was detected or was not detected, or did or did not exceed a specified concentration threshold such as a method reporting level. The number of samples required to provide a specified confidence interval for the unknown probability (p) of a VOC being detected (or exceeding some concentration threshold) with an allowable error (d) of "+ or – d" can be determined from the binomial distribution (Iman and Conover 1983). With 1,000 samples, p can be determined to be within + 3.1 percent at the 95-percent confidence level. Previous USEPA national surveys of VOCs had similar or smaller sample sizes (Westrick 1990). Also, estimates of the number of samples that could physically be processed monthly by the MWDSC laboratory, in addition to its normal regulatory sample load, were considered to ensure that the laboratory could process the samples from this survey.

Of the 54,000 CWSs in the United States, only about 11 percent rely on surface-water sources (U.S. Environmental Protection Agency 1998b). However, in 1998, these surface-water sources supplied nearly 168 million people or two-thirds of the total population served by CWSs. Most CWSs (about 89 percent) are exclusively or primarily ground-water supplied, but they collectively served just one-third of the total population, or about 84 million people in 1998 (U.S. Environmental Protection Agency 1998b). Thus, a sample size of 1,000 CWSs selected exclusively on the basis of distribution by source water would result in about 89 percent of the systems sampled in the Random Survey being ground-water sources. This would strongly bias the results of the Random Survey towards small systems in rural parts of the country. Conversely, if the design considered only the population served by CWSs, two-thirds of the sampled CWSs would be surface-water supplied, and the results of the survey again would be biased. A more balanced design for the Random Survey consequently considered both factors—source of water and population served—as stratification factors.

Data on the number of CWSs and the population served by active CWSs were obtained from the USEPA's Safe Drinking Water Information System (SDWIS) on November 5, 1998 (U.S. Environmental Protection Agency 1998b). These data indicated that there were 54,305 active CWSs serving 251,659,380 people in the United States, Native American Lands, and Puerto Rico on that date. According to the SDWIS data base, however, 7,345 systems purchased 100 percent of the water they distributed from other CWSs. Systems that purchased 100 percent of their water supplies did not have ground- or surface-water sources that could be sampled for this survey. Therefore, the 7,345 purchased-water CWSs were excluded from the count of systems, but not from data pertaining to population served, used to design the Random Survey.

The range in population served by CWSs was divided into five CWS-size categories in this survey. Generally, they were the same categories that were used by the USEPA when presenting information on the occurrence of contaminants in drinking water (U.S. Environmental Protection Agency 1993) with one difference—the category that defines the very large CWSs included all systems that serve 50,000 or more people rather than the 100,000 or more people used by the USEPA. This change was intended to include more of the larger systems, and a larger portion of the population in the random design. The five CWS-size categories used in the Random Survey are given in Table 3.3.

Table 3.3
Number of self-supplied community water systems and number of people served, by source of water and size of system, November 5, 1998

CWS-size category	Ground water Number of systems	Ground water Number of People served	Surface water Number of Systems	Surface water Number of people served
Very small	28,324	4,625,130	1,228	616,012
Small	9,775	14,178,037	1,562	5,739,217
Medium	2,399	14,219,831	971	11,045,463
Large	1,194	25,342,137	928	36,525,585
Very large	182	25,696,338	397	113,671,630
Total	41,874	84,061,473	5,086	167,597,907

Data from U.S. Environmental Protection Agency's Safe Drinking Water Information System URL http://www.epa.gov/enviro/html/ sdwis/

CWS - community water system

The design of the Random Survey distributed the total number of samples (1,000) among 10 source-size categories (five for ground water, five for surface water) relative to the national distribution of total number of self-supplied systems and the total number of people served by CWSs within each category (Tables 3.3 and 3.4). The percentage of the total number of CWSs within each source-size category (A) and the percentage of the total population served by all CWSs within that particular source-size category (B) was summed and divided by 2 (C) (Table 3.4). The number of CWSs to be sampled per category (D) was the number obtained (C) multiplied by 10. Consequently, the distribution of CWSs sampled included 311 very small ground-water systems reflecting the prevalence of such systems in the national distribution, but also included 230 very large surface-water systems—reflecting the large percentage of the population served by those systems (Table 3.4).

Table 3.4
Number of community water systems to be sampled during Random Survey on basis of mean percentage of total number of systems and total number of people served by size category and source of water

Source-size category	(A) Percentage of total number of systems	(B) Percentage of total number of people served	(C) (A+B)/2	(D) Number of CWSs to be sampled by source and size
GW-VSM	60.3	1.8	31.1	311
GW-SM	20.8	5.6	13.2	132
GW-MED	5.1	5.6	5.3	54
GW-LRG	2.5	10.1	6.3	63
GW-VLRG	0.3	10.2	5.3	53
Ground-water subtotal	89.1	33.4	61.3	613
SW-VSM	2.6	0.2	1.4	14
SW-SM	3.3	2.2	2.8	28
SW-MED	2.0	4.3	3.2	32
SW-LRG	1.9	14.5	8.2	83
SW-VLRG	0.8	45.2	23.0	230
Surface-water subtotal	10.9	66.6	38.7	387
Total	100.0	100.0	100.0	1,000

GW - ground water; SW- surface water; VSM- very small; SM- small; MED- medium; LRG-large; VLRG-very large;
CWSs- community water systems

The Random Survey planned to sample 613 ground-water-supplied systems and 387 surface-water-supplied systems. Using the mean population served by each CWS within the 10 source-size categories, a survey of 1,000 CWSs would provide information on VOC concentrations in source water used by an estimated 80 million people, or about 31 percent of the total population served by CWSs as of November 5, 1998. Using source-water type alone in the design, information pertinent to only about 2 million people would be included in the survey.

Allocation of Samples by State

In an extension of the overall design of the Random Survey, a stratified random selection of CWSs was conducted on the basis of the number of systems and population served within each of 52 geographic entities—the 50 states, Native American Lands, and Puerto Rico. The stratified random selection was made from lists of CWSs in each of the 52 geographic entities to provide a more representative distribution of participating CWSs than might be obtained from a simple random selection from the total, national population of CWSs in each category.

Similar to the overall design process, an average of the percentage of the number of CWSs and the percentage of the population served within each source-size category was calculated for each state, and for Native American Lands and Puerto Rico. This mean percentage for each of the 52 geographic entities was multiplied by the total number of CWSs to be sampled from each category in the overall design (Table 3.4). For example, in California there were 1,963 ground-water-supplied, very small CWSs that collectively supply 269,727 people (U.S. Environmental Protection Agency 1998b). Consequently, California had 6.9 percent of the 28,324 CWSs and 5.8 percent of the 4.6 million people served by CWSs in the very small category (Table 3.3). Averaging these two values, California would have 6.4 percent of the 311 ground-water-supplied, very small CWSs to be sampled in the Random Survey, or (with rounding) 20 selected CWSs from this source-size category. Table 3.5 lists the number of participating CWSs from each geographic entity for each of the 10 source-size categories. Selections were made from randomized lists of active, self-supplied water utilities obtained from the SDWIS (U.S. EPA 1998b), until the requisite number of systems was obtained for each source-size category in each state, Native American Lands, and Puerto Rico.

Lists of CWSs for each source-size category and geographic entity where one or more participating CWSs were needed (Table 3.5) were randomized using the uniform random distribution method in Microsoft Excel 5.0 (Microsoft, Inc., Seattle, Washington). Random numbers between 0 and 1.0 were generated for each CWS and sorted from lowest to highest. Identification of participating suppliers were made by contacting each CWS in sequence (from the list) to solicit their participation in the survey. This process was followed until the requisite number of systems (Table 3.5) was achieved.

Table 3.5
Community water systems to be sampled during the Random Survey by source-size category and by State

State* or other entity	Source-size category GW-VSM	GW-SM	GW-MED	GW-LRG	GW-VLRG	SW-VSM	SW-SM	SW-MED	SW-LRG	SW-VLRG	Total
AK	4	1	0	0	0	1	0	0	0	1	7
AL	0	2	2	1	0	0	0	1	3	5	14
AR	2	2	1	0	0	0	1	1	1	1	9
AZ	5	2	1	2	1	0	0	0	0	6	17
CA	20	5	3	7	11	2	2	2	5	30	87
CO	5	1	0	0	0	1	1	1	2	6	17
CT	5	1	0	0	0	0	0	0	2	5	13
DE	2	1	0	0	0	0	0	0	0	1	4
FL	13	6	3	7	17	0	0	0	1	3	50
GA	12	3	1	1	1	0	1	1	3	8	31
HI	0	1	0	0	1	0	0	0	0	0	2
IA	7	4	1	1	1	0	0	0	0	2	16
ID	5	1	1	1	0	1	0	0	0	1	10
IL	7	5	2	2	1	0	1	1	3	8	30
IN	4	4	2	2	1	0	0	1	1	4	19
KS	4	3	0	1	0	0	1	1	1	3	14
KY	1	0	0	0	0	0	1	2	5	2	11
LA	8	5	3	2	2	0	0	1	1	5	27
MA	2	1	2	3	1	0	0	1	4	8	22
MD	4	1	1	1	0	0	0	0	1	5	13
ME	2	1	0	0	0	0	1	0	1	0	5
MI	9	4	1	1	1	0	0	1	2	4	23
MN	6	4	1	3	0	0	0	0	1	2	17
MO	8	4	2	1	1	0	1	1	1	4	23
MS	5	9	3	2	0	0	0	0	0	0	19
MT	5	1	0	0	0	0	0	0	0	1	7

(continued)

Table 3.5 (continued)

State* or other entity	Source-size category										Total
	GW-VSM	GW-SM	GW-MED	GW-LRG	GW-VLRG	SW-VSM	SW-SM	SW-MED	SW-LRG	SW-VLRG	
NC	16	3	1	1	0	0	1	2	4	7	35
ND	1	1	0	0	0	0	0	0	1	0	3
NE	5	2	1	1	0	0	0	0	0	1	10
NH	5	1	0	0	0	0	0	0	1	1	8
NJ	3	2	2	4	1	0	0	0	1	8	21
NM	5	1	1	1	1	0	0	0	0	0	9
NV	2	1	0	0	0	0	0	0	0	1	4
NY	15	4	1	3	4	1	2	2	4	16	52
OH	7	4	2	3	2	0	1	1	4	10	34
OK	3	2	1	1	0	1	2	2	2	4	18
OR	6	1	0	1	0	1	1	1	2	2	15
PA	14	5	2	1	0	1	1	2	6	15	47
RI	0	0	0	0	0	0	0	0	0	1	1
SC	4	1	1	0	0	0	0	0	2	3	11
SD	2	1	0	0	0	0	0	0	0	1	4
TN	1	1	1	1	1	0	1	2	4	4	16
TX	24	14	6	3	3	1	2	2	5	17	77
UT	3	1	1	1	0	0	0	0	0	4	10
VA	11	2	0	0	0	0	1	1	2	6	23
VT	3	1	0	0	0	0	1	0	0	0	5
WA	17	4	1	2	1	1	1	0	1	2	30
WI	7	4	2	1	1	0	0	0	1	3	19
WV	2	1	0	0	0	0	2	1	1	1	8
WY	2	0	0	0	0	0	0	0	0	0	2
NA†	6	2	0	0	0	1	0	0	0	0	9
TE‡	2	1	1	1	0	2	2	1	4	8	22
Total	311	132	54	63	53	14	28	32	83	230	1,000

* Postal Service State abbreviation.

† Native American Lands.

‡ Puerto Rico.

GW, ground water; SW, surface water; VSM, very small; SM, small; MED, medium; LRG, large; VLRG, very large

Temporal Distribution of Samples

Because the Random Survey was implemented over a 15-month period, the temporal distribution of CWS sample collection also was distributed to prevent any seasonal bias in the data. Sampling of source water at selected CWSs was planned to occur from April 1999 through June 2000. Although the sampling was planned to take place over a 64-week period, the number of samples collected weekly provided for an even distribution over a 52-week calendar year. Accomplishing that distribution required that about one-half of the number of weekly samples collected during the middle 40 weeks of the survey be collected during the initial and final 12 weeks. Consequently, 10 samples per week were planned to be collected during weeks 1 through 12 and weeks 53 through 64, whereas 19 samples per week were planned to be collected during weeks 13 through 52.

A spatial distribution of CWSs to be sampled during any specific week over the 64-week duration of the Random Survey was needed to prevent any regional bias in the temporal sample distribution. Generic identification of the 1,000 CWSs sampled allowed a pre-selection and temporal randomization of the CWS sampling sequence. Information on the source-size category, state, and number of CWSs from Table 3.5 were combined to create a generic identifier for each of the planned 1,000 CWSs. For example, Table 3.5 indicates that there were four ground-water-supplied, very small CWSs sampled from Alaska; therefore, four of the generic identifiers were "GW_VSM.AK1," "GW_VSM.AK2," "GW_VSM.AK3," and "GW_VSM.AK4." Similar identifiers were created for the other 996 CWSs to be included in the Random Survey as indicated in Table 3.5. The generic identifiers served as "place holders" in the random temporal distribution design until actual selection of the participating CWSs was complete.

The list of the 1,000 generic identifiers were randomized using the uniform random distribution method in Microsoft Excel 5.0 (Microsoft, Inc., Seattle, Washington). Random numbers between 0 and 1.0 were generated for each CWS and sorted from lowest to highest. Generic identifiers then were apportioned among the 64-week sampling sequence according to the sorted random numbers and the number of samples allocated to each week (either 10 or 19). Table 3.6 gives the results of this procedure for the first and last weeks of the Random Survey sampling sequence when 10 samples were collected during each week.

Table 3.6
Community water systems to be sampled during first and last weeks of Random Survey

Sample sequence	Random number	Sample week	Generic identifier	CWS State*
1	0.002807703	Week 1	GW_SM.IN1	IN
2	0.004974517	Week 1	GW_LRG.OH1	OH
3	0.005157628	Week 1	GW_SM.MO1	MO
4	0.005920591	Week 1	GW_VSM.WI1	WI
5	0.008178961	Week 1	GW_VSM.GA1	GA
6	0.008575701	Week 1	SW_VLRG.CA1	CA
7	0.014191107	Week 1	SW_VSM.NA1	NA
8	0.014221625	Week 1	GW_VLRG.FL1	FL
9	0.014252144	Week 1	SW_VLRG.MO1	MO
10	0.014496292	Week 1	GW_LRG.CA1	CA
991	0.990447707	Week 64	GW_VLRG.CA11	CA
992	0.991241188	Week 64	GW_LRG.MA3	MA
993	0.992522965	Week 64	SW_VLRG.MI4	MI
994	0.993255409	Week 64	GW_VSM.NA6	NA
995	0.993774224	Week 64	SW_VLRG.CA30	CA
996	0.994232002	Week 64	SW_LRG.TE4	TE
997	0.995330668	Week 64	GW_VSM.MN6	MN
998	0.997711112	Week 64	GW_SM.FL6	FL
999	0.998168889	Week 64	GW_VSM.WA17	WA
1,000	0.998718223	Week 64	GW_VSM.VA11	VA

* Postal Service State abbreviation.

CWS - community water system; GW- ground water; SW - surface water; VSM - very small; SM - small; MED - medium; LRG - large; VLRG - very large

Table 3.7 includes information used to evaluate how well the random temporal sample distribution allocated samples from various parts of the nation and its Territories over a 12-month "calendar year." The 50 states, Native American Lands, and Puerto Rico were assigned to eight "regions," and the total number of samples to be collected within each region during each calendar month (assuming that week 1 of the survey was May 2-8, 1999, and week 64 was be July 16-22, 2000) were counted. For most regions, the month-to-month variability in the number of samples collected was small, and the coefficient of variation for all regions except "other" was less than 0.5 and mostly less than 0.3. The "other" region had the fewest samples (40), but it also included geographic entities that were not in similar geographic or climatic regions (Alaska, Hawaii, Native American Lands, and Puerto Rico).

Table 3.7
Community water systems to be sampled during Random Survey over a 12-month calendar year in eight regions of the United States

Region*	Jan	Feb	Mar	Apr	May	June	July	Aug	Sep	Oct	Nov	Dec	Total
Northeast and Mid-Atlantic	15	21	22	8	21	23	15	17	22	13	16	21	214
Southeast	10	11	20	13	15	8	13	11	18	11	11	16	157
Ohio Valley	12	17	7	10	13	7	10	10	11	12	8	8	125
Upper Midwest	6	7	7	12	8	9	17	6	6	9	9	10	106
South Central	11	11	11	14	10	16	19	9	13	7	10	19	150
Southwest	13	5	17	13	7	11	12	12	16	11	15	12	144
Northwest	3	3	7	5	4	4	4	8	6	7	7	6	64
Other	6	1	4	5	2	2	5	3	3	6	0	3	40
Total	76	76	95	80	80	80	95	76	95	76	76	95	1,000

* Regions (Postal Service State abbreviation):
Northeast and Mid-Atlantic: CT, DE, MA, MD, ME, NH, NJ, NY, PA, RI, VA, and VT;
Southeast: AL, FL, GA, NC, SC, and TN;
Ohio Valley: IL, IN, KY, MI, OH, and WV;
Upper Midwest: IA, KS, MN, MO, ND, NE, SD, and WI;
South Central: AR, LA, MS, OK, and TX;
Southwest: AZ, CA, CO, NM, NV, and UT;
Northwest: ID, OR, MT, WA, and WY;
Other: AK, HI, Native American Lands, and Puerto Rico.

Ancillary Information Collection

Ancillary information was collected and statistical analysis was conducted to determine possible relations between the occurrence and distribution of MTBE or other VOCs in source-water samples (collected for the Random Survey) with various natural or anthropogenic factors. Ancillary information included location (latitude and longitude) of the sources sampled (well head or intakes); actual population served by the source; source characteristics (for example, well depth, yield, aquifer type, surface-water type and size, intake specifics, and previous water-quality problems); MTBE and other fuel-oxygenate use areas; land use; population density; and known or potential VOC point-source locations (Toxic Release Inventory Sites, LUSTs, and sites regulated by the Resource Conservation and Recovery Act or Comprehensive Environmental Response and Compensation and Liability Act) in the vicinity of sampled CWS sources. This information was obtained from available data bases or collected directly from participating CWSs.

Information on the precise location of CWS sources often is missing from the SDWIS data base and was collected from the participating CWSs. National geospatial data on land use, population density, and point-source locations were available and were utilized to provide a consistent coverage for all participating CWSs. Information on MTBE or other fuel-oxygenate use was compiled from USEPA documentation and industry surveys.

Phone survey. A short questionnaire was filled out during an initial telephone interview with CWS staff when they agreed to participate in the Random Survey. The questionnaire was used to verify basic information obtained from the SDWIS data base for the selected utility, ascertain additional information on the source water to be sampled, and identify the persons and the means for further contact.

The CWSs for the Random Survey were contacted in the sequence determined from the random selection process and asked to participate in the Random Survey. System contacts were made based on the scheduled week of sampling (i.e., sampling spread out over 64 weeks). The caller described the project to the potential participant and solicited their participation. After successful agreement to participate, the caller described general sampling requirements and the actual date that the participating system would collect their source-water samples. The CWS participant was also instructed that they would receive a sampling kit approximately two weeks prior to the sample date and results approximately two months after sampling. The caller then proceeded to conduct the phone survey and entered the survey information into a Microsoft Access database. The phone survey is shown in Appendix A.

Mail In Survey. A written questionnaire (Appendix B) was included with the sampling supplies sent to participating CWSs to record additional ancillary information about the source water, intake location, filtration and treatment, distribution area, actual population served, and quantity of water delivered by suppliers. The written questionnaire also solicited information specific to ground-water sources (well characteristics, aquifer type, and any known or potential VOC point-source locations) or surface-water sources (watershed protection, watercraft use). In addition, latitude and longitude information was collected from each utility for each sampled source (intake or wellhead). If latitude and longitude were not available, the CWS was asked to identify the location of sources (wellhead or intake) on a topographic map, from which USGS personnel would determine the latitude and longitude. The latitude and longitude were entered into a geographic information system (GIS). To confirm latitude and longitude locations (if provided) or to determine the location of the source, a follow-up telephone call was made during which the CWS contact was asked to verbally describe the location of the well or intake. The source was then located, using Delorme Street Atlas/TopoUSA (Delorme, Yarmouth, Maine), and the latitude and longitude were determined. The CWSs were also asked to verbally provide information missing on the written questionnaire, if the information was available to them. This ancillary information was used to identify important natural and anthropogenic factors that may relate to the occurrence and concentration of MTBE and other VOCs observed in the source water.

The path of the mail-in survey data from source to data base was more direct. Survey forms were photocopied at MWDSC and sent to the Rapid City USGS office. Data from the survey forms were entered by hand into an Access data base. A number of possible errors could occur that may alter what the respondent intended to what ends up in the data base. These include:

1. Errors created by the respondent (e.g. misspelled street name)
2. Typographical errors in data entry
3. Interpretation errors (due to illegible handwriting of respondent)
4. Interpretation errors (data entry person must interpret response or summarize response)

Verification of the mail-in survey data involved selecting a random subset (2-3%) of survey forms and comparing all responses to what was in the data base. Errors such as #2 were easily corrected. Errors such as #4 were identified, and if a systematic misinterpretation or misrepresentation was occurring, it could be identified and remedied.

USGS staff has reviewed the data base for these systematic type errors. Several questions that have led to discrepancies between the paper form and the data base have been identified. Most of these involve yes/no type questions where the respondent provided more information than was asked for. In these instances the yes/no response was recorded, and a box was checked in an ancillary field indicating that a comment exists in the paper file. This allowed for consistency in data entry, and indicated that more information was present in the paper file.

Focused Survey

The Focused Survey was designed to determine the frequency of detection, the range in concentrations, and the temporal variation of MTBE, three other ether gasoline oxygenates, and 62 other VOCs (Tables 3.1 and 3.2) in drinking-water sources. In addition, the occurrence and range in concentrations of ether oxygenate degradation products were determined. The design of the Focused Survey was based partly on factors that were suspected or known to be related to frequent detection of MTBE and other VOCs in source water. The survey was conducted in two parts: (1) CWSs with source water suspected of having concentrations of MTBE, and (2) CWSs with source waters with known concentrations of MTBE. This design allowed sampling to begin in areas where MTBE was suspected to occur and also allowed the flexibility to redirect sampling as new water-quality data become available from, for example, the Random Survey.

The distribution of samples for the Focused Survey by type of drinking-water source and sample frequency is shown in Table 3.8. About 480 samples were planned to be collected from CWSs and distributed among drinking-water sources. Eighty wells sampled biannually, 40 reservoirs and lakes sampled quarterly, and 20 rivers and streams sampled eight times per year were planned in the Focused Survey. Samples were collected multiple times from each source water in the Focused Survey to evaluate temporal variations on MTBE and other VOC concentrations. For example, variations in MTBE and VOC concentrations are thought to be most evident for surface-water samples collected during spring and summer months (frequent watercraft use) than in those samples collected during the fall and winter months when watercraft use is infrequent. All drinking-water sources vulnerable to MTBE and other VOCs that are utilized by the CWSs were considered as candidates for the survey. Therefore, it was possible for one CWS to have samples collected from all three source-water categories listed in Table 3.8. One half of the scheduled samples were sampled for the oxygenate by-products.

Table 3.8
Number of source-water types and sampling frequency for the Focused Survey

Source water	Number of sources	Sampling frequency	Total number of samples
Ground water	80	Biannually	160
Reservoirs and lakes	40	Quarterly	160
Rivers and streams	20	Eight per year	160
Total	140		480

As part of the design of the Focused Survey, control source-water sites for MTBE were selected. The control source water sites were included in the total number of sources in Table 3.8. These were source waters in areas where there was no known use of MTBE in gasoline. Two wells, one river, and one reservoir were selected for control sites. The wells were located in selected urban areas and withdraw their water from unconfined, unconsolidated aquifer materials. The river and reservoir sources selected had infrequent watercraft use.

Community Water Supplies with Suspected Methyl tert-Butyl Ether Contamination

CWS source water located in hydrologic, climatic, and demographic settings that were suspected of being especially vulnerable to MTBE contamination were selected initially. CWS sources that met the following criteria were selected for sampling: (1) highly urbanized, high population-density settings, in areas where reformulated and oxygenated gasoline were required, especially in areas of high precipitation and high recharge rates where VOCs may be transported quickly to shallow ground water, and (2) reservoirs, lakes, and large rivers that were used by watercraft, especially in areas where MTBE was in use.

CWS sources suspected of being contaminated with MTBE initially were selected in metropolitan areas with populations greater than 250,000 and where gasoline used in the area contained MTBE. Much of the MTBE use in the United States occurs in RFG designated areas, and about 80 percent of RFG contains MTBE to achieve a specified minimum oxygen content. Recent findings for shallow ground water (less than 200-foot well depth) indicate that the use of MTBE in gasoline in RFG and OXY areas increases the probability of detecting MTBE in ground water by as much as a factor of six (P.J. Squillace, U.S. Geological Survey, written commun., 1999). It is estimated that approximately 20 million of the 50 million people who use drinking-water supplies from ground water in RFG and OXY areas use a water supply that is considered vulnerable to contamination by MTBE, other ether gasoline oxygenates, and other VOCs (P.J. Squillace, U.S. Geological Survey, written commun., 1999). The vulnerability of aquifers to MTBE contamination appears to be related mostly to use of MTBE in gasoline, the density of above-ground and underground storage tanks, and soil erodibility (Squillace and Moran 2000). Other factors, such as well depth and depth to ground water, and the presence of roads seem to be less important.

A total of 139 of the largest cities and metropolitan areas in the United States were identified by USGS from 1992 census data projections. Of the 139 largest cities and metropolitan areas, 27 were determined to be in RFG areas or in the "opted-in" RFG program (U.S. Environmental Protection Agency 1999a). Thirteen of the 139 cities and metropolitan areas were determined to be in OXY gasoline areas where ethanol is the primary oxygenate in gasoline (U.S. Environmental Protection Agency 1999b).

The Motor Gas Survey, which provides chemical analyses of gasoline collected from 111 select metropolitan locations across the United States, has included ether oxygenate analyses by an independent firm (TRW Systems and Information Technology Group, Bartlesville, Oklahoma) since about 1991. The most recent data from this survey, collected from 1996 and 1997, were used to compare MTBE concentrations in gasoline in the 139 large cities and metropolitan areas. Where MTBE was detected in gasoline, the concentrations were either less than 1 percent by volume, or approximately 10 percent by volume. Metropolitan areas using RFG gasoline were found to have the highest MTBE concentrations in gasoline. Lower concentrations of MTBE likely are due to its use as an octane enhancer.

On the basis of population density, federally designated RFG/OXY areas, and MTBE data from gasoline analyses, 14 metropolitan areas were selected initially for the Focused Survey. All selected metropolitan areas had populations greater than 250,000, with MTBE concentrations in gasoline equal to or greater than 10 percent by volume as determined from the Motor Gas Survey (Table 3.9). About 2.6 percent of all CWSs located in the 14 metropolitan areas fall into the very large source-size category (more than 50,000 people per system). These CWSs provide water to more than 75 percent of the population served within those 14 metropolitan areas. These were the CWSs chosen initially. CWSs that purchase water from a wholesaler were not considered for this survey. Table 3.10 lists the type of sources to be sampled in each metropolitan area. The number of sampling sites was distributed among the 14 metropolitan areas on the basis of population. Therefore, of the 14 metropolitan areas, the more populated areas had more sampling sites in comparison to less populated areas.

Table 3.9
Metropolitan areas where methyl *tert*-butyl ether is used extensively in gasoline

Identifier	Metropolitan area
1	Bakersfield, CA (MSA)
2	Boston, MA (NECMA)
3	Dallas-Ft. Worth, TX (CMSA)
4	Hartford, CT (NECMA)
5	Houston-Galveston-Brazoria, TX (CMSA)
6	Los Angeles-Riverside-Orange County, CA (CMSA)
7	Louisville, KY-IN (MSA)
8	New York-Northern New Jersey-Long Island, NY-NJ (CMSA)
9	Philadelphia-Wilmington-Atlantic City, PA-NJ-DE-MD (CMSA)
10	Providence-Fall River-Warwick, RI-MA (NECMA)
11	Richmond-Petersburg, VA (MSA)
12	San Diego, CA (MSA)
13	San Francisco-Oakland-San Jose, CA (CMSA)
14	Washington-Baltimore, DC-MD-VA-WV (CMSA)

MSA - Metropolitan Statistical Area; NECMA - New England County Metropolitan Area; CMSA - Consolidated Metropolitan Statistical Area

Some states do not have RFG or OXY gasoline program areas; however, these states may have areas that have a high detection frequency of MTBE in ground water at LUST and gasoline spill sites. Discussions with underground-storage-tank and drinking-water officials, water suppliers, and USGS personnel helped gain a better understanding of MTBE releases to source water in these states. Additional candidate aquifers, streams, and rivers for the Focused Survey were expected to be identified through these discussions. Some reservoirs and lakes that have frequent recreational use in former RFG areas or near (within 100 miles) current RFG areas were suspected of having detectable MTBE concentrations. The number and national distribution of these sources are listed in Table 3.11 by state and source-size categories. The source-size categories are identical to those used in the Random Survey (for example, the very large category is greater than 50,000 served).

Table 3.10
Number of community water sources to be sampled in 14 metropolitan areas suspected to have concentrations of methyl tert-butyl ether in source waters

Identifier	Metropolitan area (table 3.9)	Number of ground-water sources	Number of river/stream sources	Number of reservoir/lake sources
1	Bakersfield	2	1	1
2	Boston	2	1	1
3	Dallas	2	1	1
4	Hartford	2	1	1
5	Houston	2	1	1
6	Los Angeles	4	1	2
7	Louisville	2	1	1
8	New York	4	1	2
9	Philadelphia	4	1	2
10	Providence	2	1	1
11	Richmond	2	1	1
12	San Diego	2	1	1
13	San Francisco	4	1	2
14	Washington, D.C.	4	1	2
	Totals	38	14	19

Community Water Systems with Known Methyl tert-Butyl Ether Contamination

This part of the Focused Survey concerns water sources with known MTBE occurrences. CWSs were selected on the basis of the following criteria: (1) MTBE was known to occur in the source water on the basis of recently completed or ongoing sampling activities, and (2) verbal communication from local, state, or other federal agency personnel indicated the presence of MTBE in the source water. Source-water samples analyzed from the Random Survey that were determined to contain concentrations of MTBE greater than or equal to 0.5 μg/L were included in the Focused Survey to expand the amount of data pertaining to contaminated source water. In addition, information gained from a 12-state drinking-water retrospective (Grady and Casey 2001), recent reports completed by the states of Maine and California (Keller et al. 1998), and other ancillary information (Chapter 2) as it became available were used to select CWSs to be included in the Focused Survey. Table 3.12 provides a summary of the states and source-size categories of CWSs with sources having known occurrences of MTBE.

Table 3.11
Number of community drinking-water sources suspected of having concentrations of methyl *tert*-butyl ether by State and source-size category

State*	GW-VSM	GW-SM	GW-MED	GW-LRG	GW-VLRG	SW-VSM	SW-SM	SW-MED	SW-LRG	SW-VLRG	Total
	Source-size category										
CA	0	0	2	4	7	0	0	0	1	6	20
CO	0	0	0	0	0	0	0	0	0	1	1
CT	0	0	0	2	1	0	1	0	0	0	4
DE	2	2	3	0	1	0	0	0	0	2	10
GA	0	0	0	1	0	0	0	0	0	2	3
IN	0	0	0	0	0	0	0	0	0	1	1
KY	0	0	0	2	0	0	0	0	0	1	3
LA	0	0	0	0	0	0	0	0	0	1	1
MA	0	0	0	0	0	0	0	0	0	2	2
MD	0	0	0	0	0	0	0	0	0	1	1
MS	0	0	0	0	0	0	0	0	0	1	1
NJ	0	0	0	0	0	0	0	0	0	1	1
NV	0	0	0	0	0	0	0	0	0	1	1
NY	0	0	0	0	4	0	0	0	0	0	4
OK	0	0	0	0	0	0	0	0	0	2	2
OR	0	0	1	3	0	0	1	1	0	0	6
PA	0	0	0	0	0	0	0	0	0	3	3
RI	0	0	0	0	0	0	0	0	0	1	1
TX	0	1	0	0	2	0	0	0	0	3	6
VA	0	0	0	0	0	0	0	0	0	3	3
VT	0	0	0	0	0	0	0	0	0	1	1
WV	0	0	2	0	0	0	0	0	0	0	2
Total	2	3	8	12	14	0	2	1	1	33	77

* Postal Service State abbreviation.

GW - ground water; SW - surface water; VSM - very small; SM - small; MED - medium; LRG - large; VLRG - very large

Table 3.12
Number of community drinking-water sources with known concentrations of methyl *tert*-butyl ether by State and source-size category

State*	Source-size category GW-VSM	GW-SM	GW-MED	GW-LRG	GW-VLRG	SW-VSM	SW-SM	SW-MED	SW-LRG	SW-VLRG	Total
AZ	0	0	0	1	0	0	0	0	0	1	2
CA	0	0	0	0	0	0	1	0	0	8	9
CT	0	1	0	0	0	0	0	0	0	0	1
DE	0	0	0	1	0	0	0	0	0	0	1
FL	0	0	0	0	2	0	0	0	0	1	3
IA	1	1	0	0	0	0	0	0	0	0	2
IL	0	2	2	0	0	0	0	0	0	0	4
KY	1	0	0	0	0	0	0	0	0	0	1
KS	0	2	1	1	0	0	0	0	0	0	4
MA	1	0	0	1	0	0	0	0	0	0	2
MD	0	1	1	1	0	0	0	0	0	0	3
ME	0	1	2	1	0	0	0	0	0	0	4
NC	0	0	1	0	0	0	0	0	0	1	2
NH	0	0	1	1	0	0	0	0	1	1	4
NJ	0	0	0	1	3	0	0	0	0	0	4
NY	0	0	0	0	1	0	0	0	0	0	1
OH	0	1	0	0	0	0	0	0	0	0	1
OR	2	0	0	0	0	0	0	0	0	0	2
PA	1	0	0	0	0	0	0	0	0	1	2
TN	0	0	0	0	0	0	0	0	1	0	1
TX	0	0	0	0	0	0	1	0	0	2	3
VA	0	1	0	0	0	0	0	0	0	0	1
Total	6	10	8	8	6	0	2	0	2	15	57

* Postal Service State abbreviation.

GW - ground water; SW - surface water; VSM - very small; SM - small; MED - medium; LRG - large; VLRG - very large

Ancillary Information Collection

As with the Random Survey, ancillary information on each CWS and source water was collected for the Focused Survey. The information was used for spatial and temporal analysis and to determine possible relations between the occurrence of MTBE in the source-water samples and anthropogenic factors. Similar ancillary information for the Focused Survey as for the Random Survey was sought: latitude/ longitude of the sources sampled, population served, source characteristics, MTBE and other gasoline oxygenate use data and designated program areas, and known or potential VOC point-source locations.

A short telephone questionnaire was completed during the initial interview with the participating CWS staff to identify the contact person and get additional information on the source water (Appendix C). A written, more detailed questionnaire (Appendix D) was delivered to the CWS contact by the USGS sampling personnel. The written questionnaire for the Focused Survey requested similar information (intake/well location, water treatment, any known or potential VOC point-source locations, well characteristics, watershed protection) as the written questionnaire for the Random Survey. The USGS sampling personnel provided the location (latitude/longitude) of the source water either in the form of a point on a topographic map or global positioning system (GPS) coordinates.

FIELD METHODS - SAMPLING

Sample collection technique is critical to the validity of the analytical data. Quality assurance procedures and safeguards must be adhered to in order to ensure the integrity of the data. The potential exists for both contamination of the sample as well as loss of VOCs due to improper sampling technique.

Volatile compounds in the environment at a sample site can contaminate a sample during collection. Field reagent blank (FRB) samples were used to measure any contamination from the environment around the sample site and during the sample shipment. FRBs were prepared at each sample site to measure any background contamination at the site. A bottle containing organic free water was provided with each sample kit. Commercially produced organic free blank water (J.T. Baker, Phillipsburg, N.J.) was randomly tested at the MWDSC laboratory to ensure purity and sent to the CWSs to be used for collection of FRBs. Empty VOC sample vials labeled "**Field Blanks**" were opened at the sample site and filled with the organic free water provided in the sample kit. The vials were sealed headspace free as described in the sampling instructions and shipped back to the laboratory with the samples. The field blanks were analyzed only if VOCs were detected in the corresponding environmental sample. Field equipment blanks (FEBs) were collected as described by Shelton (1997).

Trip reagent blanks (TRBs), filled with organic free water and sealed at the MWDSC, also accompanied all environmental samples. TRBs were never uncapped by the samplers but were analyzed if VOCs were detected in the environmental sample.

The sample collectors were provided with a warning to ensure that the sample site was free of environmental contamination. It was recommended that the area within 100 feet of the sample site be free of any possible VOC contamination (e.g., fuel driven pumps, motors, chemical or fuel storage, etc.). The sample collectors were also instructed to keep vehicles a minimum of 100 feet away (with engines off) from the sample site. However, the extent to which water utility staff followed these instructions is unknown.

Sample Containers, Preservative and Shipment

All the sampling kits (including sample vials, blank water, instructions) were prepared at the MWDSC laboratory. The VOC samples were collected in 40-mL glass vials with PTFE-lined septa sealed open caps. The glass vials were washed with detergent and rinsed with tap water, followed by deionized water and organic free water. The vials were then baked in an annealing oven at 400°C for one hour. The caps and septa were detergent washed and rinsed with tap water, deionized water, organic-free water and acetone. The caps and septa were then baked at 180°C for one hour.

The VOC vials for the Random Survey contained 2 drops of 50 % hydrochloric acid (VWR Scientific, San Dimas, Calif.) for sample preservation through pH adjustment. VOC vials for the Focused Survey contained no dechlorination agent and 2 drops of 50% hydrochloric acid were added to the samples during collection. Samples were stored on blue ice and shipped via overnight express to the MWDSC. Samples for VOCs were analyzed within 14 days as prescribed in USEPA Method 524.2. Samples and blanks analyzed by the OHSU laboratory were collected using the same USGS field protocols; the only variation was no addition of hydrochloric acid. Samples for the oxygenate by-products were chilled only and acid preservation was avoided to prevent acid corrosion of the mass spectrometer source and damage to the gas chromatograph's columns (Appendix E).

Sample kits were shipped to each utility (or USGS field personnel for the Focused Survey sampling) approximately two weeks prior to the scheduled sample collection date. A return shipping label and prepaid courier bill were included for return of samples. The sample kit included: insulated shipping container, blue ice, VOC vials, oxygenate by-product vials (Focused Survey only), blank water for FRBs, TRBs, and sample instruction sheet (Figure 3.1).

Sample Collection

The point of contact person from each CWS conducted sample collection (environmental samples, and FRBs) for the Random Survey. The Focused Survey sampling was conducted by USGS personnel. The USGS field personnel utilized USGS sampling protocols and ground-water samples were collected from water-supply wells as described by Koterba, Wilde, and Lapham (1995).

Many of the CWSs sampled were from the small to very small categories and the level of sampling expertise ranged from very limited to fully experienced. Therefore the sampling instructions for the Random Survey were designed to be simple and direct. The sample collectors (CWS operators) for the Random Survey were instructed to: 1) collect the sample from an existing sample tap, conduit or valve (if one existed) and 2) to be sure that the water was representative of the water entering the treatment facility (if treatment was provided). The sample collectors for the Random Survey were instructed to allow the intake tap to run for at least 5 minutes prior to collection of the sample. The sampling during the Focused Survey was conducted by USGS personnel, using standard field protocols for the collection of VOC ground and surface water samples. Groundwater samples were collected from a tap prior to any treatment (including chlorination). Surface-water samples were collected near the intake and within 5 feet of the surface using a VOC hand-sampler as described by Shelton (1997) and Halde, Delzer, and Zogorski (1998).

Samples were collected in triplicate at each location. Sample bottles were filled with no headspace. The sample collectors for the Random Survey were also advised to use a tap or connection that had a flow adjustment, so that flow would not be too forceful to overfill and back splash the 40-mL vials.

Sample collectors for both surveys were asked to collect well samples only after the well had been pumped sufficiently to insure the water was representative of the ground water source.

I. Prior to day of sample collection

A. Place blue ice in freezer.

II. ON DAY OF SAMPLE COLLECTION

Remove blue ice from freezer and place in shipping container with sample bottles.
Contact either courier (***prepaid airbills for both are enclosed)*** on the day of sampling for pickup of samples that day.

FEDERAL EXPRESS (800-GO-FEDEX or 800-463-3339) OR
UPS (800-782-7892) (UPS will pick up from any location)

Do not sample on Fridays, samples can not be received on Saturday.

III. SAMPLE COLLECTION

1. **Be sure site area is free of contamination from volatile organics (e.g., fuel driven motors or pumps, fuel or chemical storage, industrial effluent discharge, etc.) Turn off engines of any vehicles within 50 feet of sample area.**
2. **Do not rinse bottles.** All sample collection bottles contain acid preservative.
3. **Warning ! Avoid contact.** Sample collection bottles contain two drops of 50% hydrochloric acid.
4. Completely fill the two bottles marked "FIELD BLANK" with the water provided in the kit. Place cap and liner (with white smooth side down towards the water) and seal bottles with no air space.
5. The sample tap location should represent the raw source water, prior to treatment.
6. Allow the water tap to flush for at least 5 minutes before collecting samples. If sampling from a well, allow water to flush for a minimum of 3 times the casing (or well bore) volume.
7. Adjust flow of tap so as not to overfill or back splash during sample collection. Completely fill the three bottles marked with the sample location.
8. Do not overfill the bottles, but leave no airspace (see diagram, next page).
9. Cap and seal bottles, making certain that the Teflon side (white smooth side) of the cap liner is facing down towards the water.
10. Place the sample bottles and field blanks together in the shipping container packed with the frozen blue ice.
11. Fill out the attached "**COLLECTION LIST**" with the actual date and time of sampling (the pre-printed sheet may not correspond to the correct sample date – this information will be corrected when samples are received at Metropolitan). You **do not** need to fill in the temperature, or the columns marked "TCLR" and "FCLR".
12. Deliver the shipping container to Federal Express, or have them pick it up. Use the Federal Express return labels and prepaid courier bill enclosed in the shipping container for priority overnight delivery.

Figure 3.1 Sample Collection Instructions

Generally, an acceptable purge volume equal to three times the casing (or well bore) volume was recommended. Sample collectors were advised to flush water lines for at least 5 minutes prior to collection of the samples.

A sample collection form (Figure 3.2) accompanied the samples and was filled out by the sampler. Other sample information was also collected (e.g., date/time of sample collection) and documented on the sample collection form.

Quality Assurance/quality control QA/QC samples included FRBs, TRBs, field duplicates, and FEBs. The Random Survey QA/QC samples included only FRBs and TRBs. The Focused Survey QA/QC sampling included FEBs (20 percent of the samples), field duplicates (10 percent of the samples), FRBs and TRBs (FRBs and TRBS were collected at each site).

ANALYTICAL

All samples for both the Random and Focused Surveys were analyzed for MTBE, three other ether gasoline oxygenates, and 62 other VOCs (Table 3.1) at the MWDSC laboratory in La Verne, California, using the USEPA approved Method 524.2 (U.S. Environmental Protection Agency 1992b). This method utilizes purge and trap sample concentration and detection from a gas chromatograph/mass spectrometer (GCMS). Additional analytes were added to the method and validated through the analysis of a MDL as described in the method. The MWDSC standard operating procedure is shown in Appendix F. The list of analytes with their respective MDLs and MRLs is shown in Table 3.2.

One-half of the samples collected for the Focused Survey were randomly selected to have a sequential sample collected for oxygenate by-products (e.g. one of the biannual groundwater samples would also have a sequential samples collected for oxygenate by-products) using the analytical method developed by the OHSU (Pankow and others, 1998). This method can be used for the analysis of gasoline oxygenates and their by-products (MTBE, ETBE, DIPE, TAME, TBA, methyl acetate, and tert-amyl alcohol) at microgram per liter concentration levels (Appendix E). The OHSU laboratory is one of the few laboratories in the country that has a rapid and sensitive method (low reporting limit) for analyzing oxygenate degradation products.

COLLECTION LIST: M4500 ROUTE: VSSW DATE: 05-JUL-99

SITE LOCATION	SAMPLE ID	SAMPLE TYPE	TIME COLL	TEMP °C	TCLR mg/L	FCLR mg/L	# REC'D
USA RIVER US0100001	B4500-1	VOC	____	____	____	____	3
FIELD BLANK US0100001	B4500-2	VOC	____	____	____	____	2

Notes: ____

____

Collected by:
EMP. #:________ SIGNATURE:________________________________ DATE:____________

Received by:
EMP. #:________ SIGNATURE:______________________ DATE:________ TIME:________

Data entered by:
EMP. #:________ SIGNATURE:______________________ DATE:________ TIME:________

Figure 3.2 LIMS Sample Collection List

The OHSU laboratory used a modification of the USEPA Method 524.2. It involves a heated purge during which the compounds are absorbed on a trap, then are desorbed and analyzed by GCMS. The OHSU standard operating procedure (SOP) for the heated purge and trap gas chromatograph/mass spectrometry (HP&T-GCMS) procedure is shown in Appendix E. The list of analytes with their respective MDLs, and calibration range are shown in Table 3.13.

Table 3.13
Oxygenates and Oxygenate By-Products analyzed by OHSU Laboratory

	MDL (µg/L)	Calibration Range (µg/L)
TBA	0.130	0.2 – 2.0
Methyl acetate	0.141	0.2 – 2.0
MTBE	0.078	0.2 – 2.0
DIPE	0.104	0.2 – 2.0
ETBE	0.101	0.2 – 2.0
Tert amyl alcohol	0.115	0.2 – 2.0
TAME	0.098	0.2 – 2.0

CHAPTER 4
QUALITY ASSURANCE/QUALITY CONTROL

DATA QUALITY OBJECTIVES

Evaluation of compounds detected in the environmental samples required evaluation against a series of quality assurance /quality control (QA/QC) strategies to ensure the accurate and precise reporting of the results. A strategy of reporting results has been developed by USGS (Connor et al. 1998), and modified here, to reduce the censoring of positive VOC detections at low concentrations. This strategy involved the evaluation of the data with the MDL, MRL and blanks (laboratory reagent blank, field reagent blank, field equipment blank, and trip reagent blank) results. Analytical results also required QA/QC review to assure that data were accurately reported.

The data verification strategy involved the complete path taken by the data (from the analysis instrument to the results database). First, data were electronically transferred from the GCMS data system to the MWDSC laboratory information management system (LIMS). Data in the LIMS underwent editing in accordance with the strategy for determining non detects and "estimated" values based on the MRL and MDL for each compound (discussed in detection level section below). Results were then queried from the LIMS using Microsoft Access, and the Microsoft Access data base file was transported as an email attachment to USGS and OHSU.

Verification procedures were required at all steps to ensure that the data transferred accurately and without errors. Hand entered or edited data were always verified by a second analyst to eliminate transcription errors. Procedures for verification of the data along its electronic path from one system to another were also in-place to ensure the integrity of the data.

Detection Levels

The evaluation of data at low concentrations was first conducted based on the MDL and MRL. The MRL for this study was set at the lowest calibration standard (i.e., 0.2 µg/L for all VOCs except methyl ethyl ketone [MEK] which was 2.0 µg/L). The MDL as defined (U.S. Environmental Protection Agency 1992a) is a statistical measurement of the minimum concentration of a substance that can be identified, measured and reported within 99 percent confidence that the compound is greater than zero. The MDLs were calculated by the MWDSC from seven replicates of low level standards (USEPA 1992a) and are shown in Table 3.2. The MDL values are approximately one order of magnitude less than the MRLs.

Calibration

Instrument calibration was necessary to ensure the accuracy of the analysis. The instrument was calibrated every two weeks. The GCMS calibration range was 0.2 to 10 µg/L for all VOCs, except MEK, which was 2.0 to 100 µg/L. Verification of calibration was conducted at the beginning of each day of analysis and at the end of the analysis batch (Appendix G). On several days a check standard was also analyzed in the middle of the batch of samples.

QA/QC Blanks

Positive identification of compounds detected in the environmental samples also required evaluation against a variety of blanks. Blanks were used to evaluate any background contamination and minimize the reporting of false positives. Several types of blank samples were analyzed in this project: instrument blanks (IB), laboratory reagent blanks (LRB), field reagent blanks (FRB), field equipment blanks (FEB), and trip reagent blanks (TRB). IBs were used to verify the background contributed by the analytical instrumentation (i.e., purge and trap system and GCMS). The LRBs were used to evaluate the overall contribution of background from the complete analysis. This background included the organic pure water used for the LRB, calibration standards, FRB, FEB; reagents; and laboratory glassware. The FRBs were used to evaluate the background contributed by the environment (particularly air) at the sample site during the sampling process. The FEBs were used (in the Focused Survey only) to evaluate the background contributed by the field sampling equipment. Field sampling equipment was used during the sampling of surface water sources in the Focused Survey whereas ground-water sources were sampled with permanently installed pumps. TRBs were prepared in the laboratory, traveled along with the samples but were not opened until time of analysis (at which time the septum was punctured by the autosampler/purge and trap system). FRBs, FEBs and TRBs were analyzed only if there was a detection in the corresponding source-water sample. LRBs and IBs were analyzed with each analytical batch of samples.

The organic pure water was purchased from J.T. Baker (Phillipsburg, N.J.). Each bottle was evaluated for contamination, and if all contaminant levels were acceptable (i.e., all analytes were less than the MDL, except for methylene chloride which was less than 0.2 µg/L), then the bottle was approved for use and stored in an area free of contamination. The organic pure water was then approved for use for all the blanks and the calibration standards. The bottle and lot analysis were tracked and the results of this QC analysis were correlated with the corresponding environmental samples and the different blanks.

The LRB assessed the overall background contamination from the analysis conducted in the laboratory to eliminate the reporting of false positives. It included measurement of any background contributed by the organic pure water (used for the blanks, and calibration standards), laboratory air and the instrumentation (i.e., purge and trap and GCMS system). As stated in the SOP (Appendix F), the LRBs were analyzed at the beginning of each batch, after the analysis of samples with VOCs (e.g., calibration check standards, and fortified samples), and at the end of each batch.

A daily IB was analyzed to evaluate the background contamination contributed by the analytical equipment. The IB strictly isolated the background contributed by the analytical instrumentation from other laboratory contributions (i.e., laboratory air and the laboratory reagent water). The analytical instrumentation consisted of the purge and trap unit, purge and trap autosampler and GCMS (see section 6.0 of Appendix F). The autosampler used for this project allowed for samples to be analyzed without ever being exposed to the laboratory environment. Each sample vial was pierced through the septum with a stainless steel needle by the autosampler at the time of analysis. The sample vial was then pressurized with helium to fill a 25-mL sample loop with sample, which was then transferred to the sparge vessel. The internal standard and surrogate solution was then transferred to the water in the sparger, and the purging commenced. For the IB an empty 40-mL sample was loaded in the autosampler and was processed in the same manner as actual samples, including the addition of internal standard and surrogates to the sparge vessel. The IB was analyzed with each batch of samples.

The evaluation of the IB showed the contribution from the instrumentation and when subtracted from the LRB, the result can be used to evaluate the background from the laboratory air and the laboratory reagent blank water. Since the environmental samples were not exposed to the laboratory air during analysis, positive detections in the environmental samples were compared to the IB not the LRB. Overall, the IBs were considered very clean and demonstrated that the laboratory analytical procedure contributed little, if any, background to the environmental samples (Appendix G). The background contamination in the LRBs was also very low and negligible (Appendix G).

Field Precision Samples – (Focused Survey only)

Field duplicate samples were used to evaluate the overall precision for each analyte from the time the sample was collected through the analytical procedure. Field duplicates were collected with approximately 15 percent of the samples. The precision of the field duplicates was very good (Appendix G).

Fortified Samples (Laboratory Matrix Spikes)

Ongoing evaluation of the accuracy of the analytical process in the sample matrix was performed with each batch of samples. On each day of analysis 10 percent of the samples were chosen and fortified with known quantities of the VOCs. The fortified samples (matrix spikes) represented a measure of the ability of the analytical procedure to recover the compounds of interest from a true matrix. The fortification solution was prepared from an independent source from the calibration standards and therefore, also served as an independent verification of the calibration. The USEPA Method 524.2 recognizes variability due to matrix interferences and therefore, acceptance limits were not established. Analyses were not rejected because of the results of the fortified samples.

The percent recovery for the matrix spike samples was calculated as follows:

Percent recovery = $C_{ms} - C_{es}/C_s \times 100$
C_{ms} - concentration in the matrix spike sample
C_{es} - concentration in the environmental sample
C_s – concentration spiked into the environmental sample

Overall, the recoveries from the environmental matrixes was very good and demonstrated the compounds were consistently recovered from the matrix samples at a high level of accuracy (Appendix G).

Laboratory Precision Samples

Laboratory duplicate analysis of the environmental samples allowed evaluation of the precision of the method. On each day of analysis, 10 percent of the samples were analyzed in duplicate to measure the analytical precision. On most occasions a duplicate of the matrix spike sample was prepared as the laboratory duplicate, because in most cases the compounds were not present in the environmental samples. However, if VOCs were known to be present in a particular environmental sample, then that sample was sometimes chosen as a laboratory duplicate. Therefore, there are some precision data from matrix spike samples prepared in duplicate and other precision data from duplicates of environmental samples. The results are expressed as the relative percent difference (RPD).

RPD = [Absolute Difference (Rep 1 – Rep 2)/Average (Rep 1 and Rep 2)] x 100
Rep 1 – analytical result for replicate # 1
Rep 2 – analytical result for replicate # 2

Overall, the precision of the analysis, as measured by the matrix spike duplicates and the laboratory duplicates, was very good (Appendix G).

Inter/Intra-Laboratory Verification

The accuracy of the analytical process was evaluated through internal analysis of quality control samples, performance evaluation (PE) samples (used for state accreditation), and inter-laboratory QC samples. Independent verification of the calibration was accomplished through the analysis of QC samples purchased from Ultra Scientific (North Kingstown, RI). PE samples were provided through the State of California Environmental Laboratory Accreditation Program in association with the USEPA Water Supply (WS) Series PE program for volatile organic compound analysis in drinking water. The inter-laboratory QC samples were purchased from Environmental Resources Associates (ERA). All of these demonstrated a high level of accuracy (see Appendix G).

Initial Data Validation

The data validation screening involved evaluation of the data for the QC requirements of the method, screening for background contamination and then conversion to a format for transfer to the USGS database. The GCMS software quantified each sample analysis. The raw GCMS data were evaluated against the daily QC requirements (Appendix F). In addition, the analyst evaluated the data for background contamination, which included evaluation against the MDL, the instrument blank and lowest calibration standard (Figure 4.1). After the evaluation as shown in Figure 4.1, the analyst electronically transferred the data into the MWDSC LIMS. The GCMS system transferred all values greater than the MDL to LIMS. There were three reporting fields in LIMS for each compound: 1)"qualifier", 2) text, and 3) numeric "result". The text field defaulted to not detected (ND) if there was nothing in the numeric field. Another level of data evaluation was conducted after the data were transferred to LIMS. The LIMS data were evaluated and edited to convert the three LIMS fields of data to two fields (a number field and a qualifier field) for transfer to the USGS database (Figure 4.2).

The screening for background contamination from the laboratory was conducted directly after the GCMS analysis. When a detection of a compound was found in the environmental sample, the result for that compound in the IB was evaluated. If that compound was not detected in the IB, then the detection was considered valid (provided the value was greater than the MDL). If that compound was detected in the IB, then the concentration of the compound in the environmental sample was evaluated against the concentration found in the IB (Figure 4.1). If the concentration of the compound detected in the environmental sample was equal to or less than the sum of the concentration in the IB plus the MDL, then the value of environmental sample was considered not detected. Not detected results were reported as “less than the non-detect value (NDV)”. The NDV was statistically defined to limit the chance of false positives and false negatives at less than 1 percent each (Connor et al. 1998). The NDV was equal to 2 x MDL. If the sample concentration was less than 5 times the IB or less than the lowest calibration standard, then the result was quantified and reported (in the result field) and an “E” value was placed in the qualifier field, to indicate the result was an estimate. If the sample concentration was greater than 5 x IB and greater than the MRL (i.e. greater than the lowest calibration standard), then the result was reported at the resulting concentration with no qualifier. An example of the data strategy presented in Figure 4.1 is shown in Table 4.1. Several scenarios of benzene detection in the environmental sample are presented. The MDL for benzene is 0.029 μg/L and the low calibration standard (i.e., MRL) is 0.2 μg/L. The NDV is 0.058 μg/L.

On occasion, the data were not transferred electronically to LIMS and hand entry into LIMS was required. However, all hand entered data or data edited after electronic transfer was always checked by a second analyst. The electronic transfer to LIMS was also checked for accuracy. The data in LIMS were reviewed to verify that the transfer was accurate and whether each analyte detected was a good spectral match. Any changes were edited and recorded by LIMS. After the data were reviewed in LIMS, the status for each sample number was changed to approved. The approved data were then queried from LIMS and stored in a Microsoft Access database. The data were then periodically transferred to USGS and reports were generated and sent to each CWS.

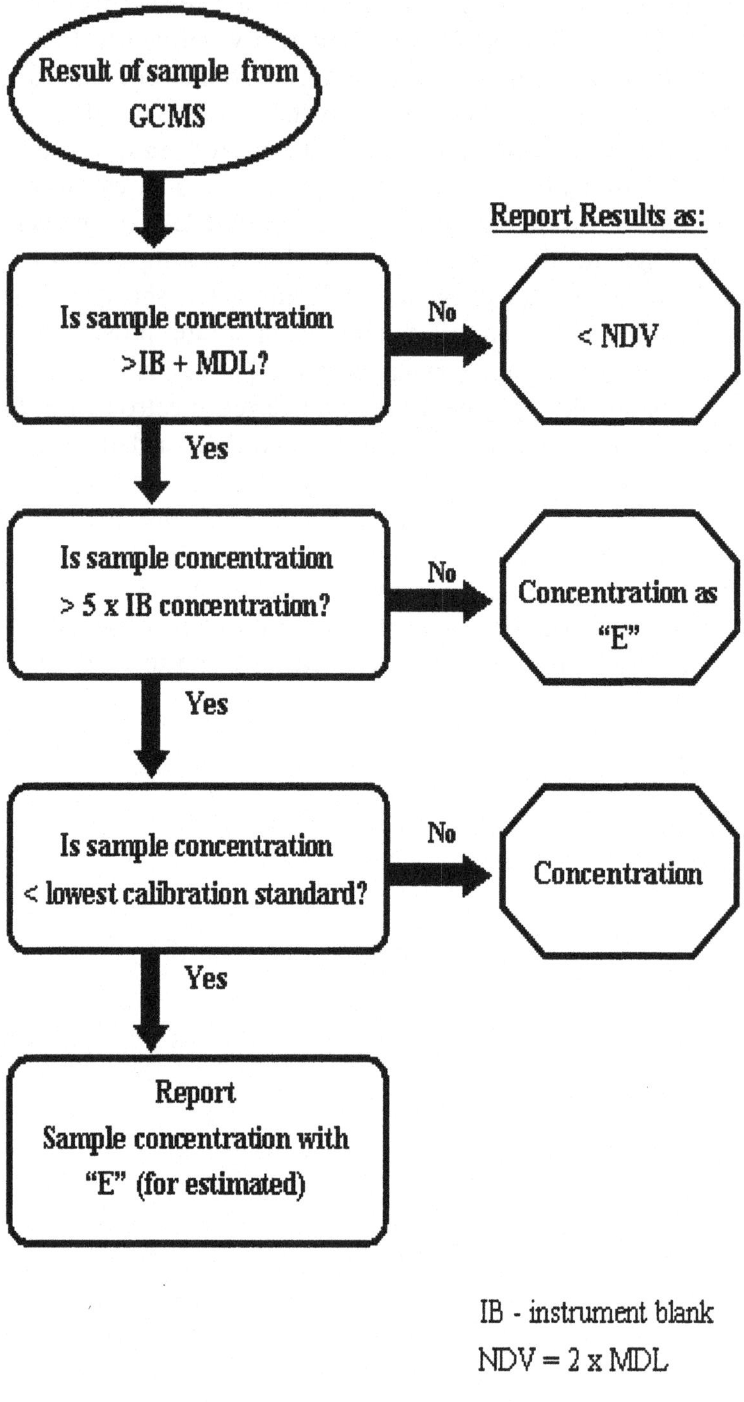

Figure 4.1 Qualification Strategy for Low Level Detections - Instrument Blank (IB) Evaluation

Table 4.1
Example scenarios of benzene data censoring strategy

IB	IB+MDL	5 x IB	Value Reported by GCMS data system	Final results reported to USGS database	
				Q Value	Result Value (μg/L)
0.01	0.039	0.05	0.015	<	0.058
0.01	0.039	0.05	0.044	E	0.044
0.01	0.039	0.05	0.06	E	0.06
0.01	0.039	0.05	0.22		0.22

IB – instrument blank
MDL – method detection limit = 0.029 μg/L
Q- qualifier field value
GCMS – gas chromatograph/mass spectrometer
Low standard (MRL) = 0.20 μg/L
NDV = 2 x MDL = 0.058 μg/L

In order to report the results to USGS, the data in the Access database were censored against the MDL and MRL according to the strategy in Figure 4.2. Two fields were reported to USGS for each compound: 1) qualifier and 2) result. All results less than the MDL were censored and reported as "< NDV" and all results less than the MRL and greater than the MDL were reported along with an "E" (for estimate) in the qualifier field. All non-detects were reported with "<" in the qualifier field and the NDV in the result field. Because many of the compound results queried from LIMS did not have corresponding results in the blank samples, data evaluation required only comparison against the MRL and MDL prior to being sent to USGS for inclusion in the project database (Figure 4.2).

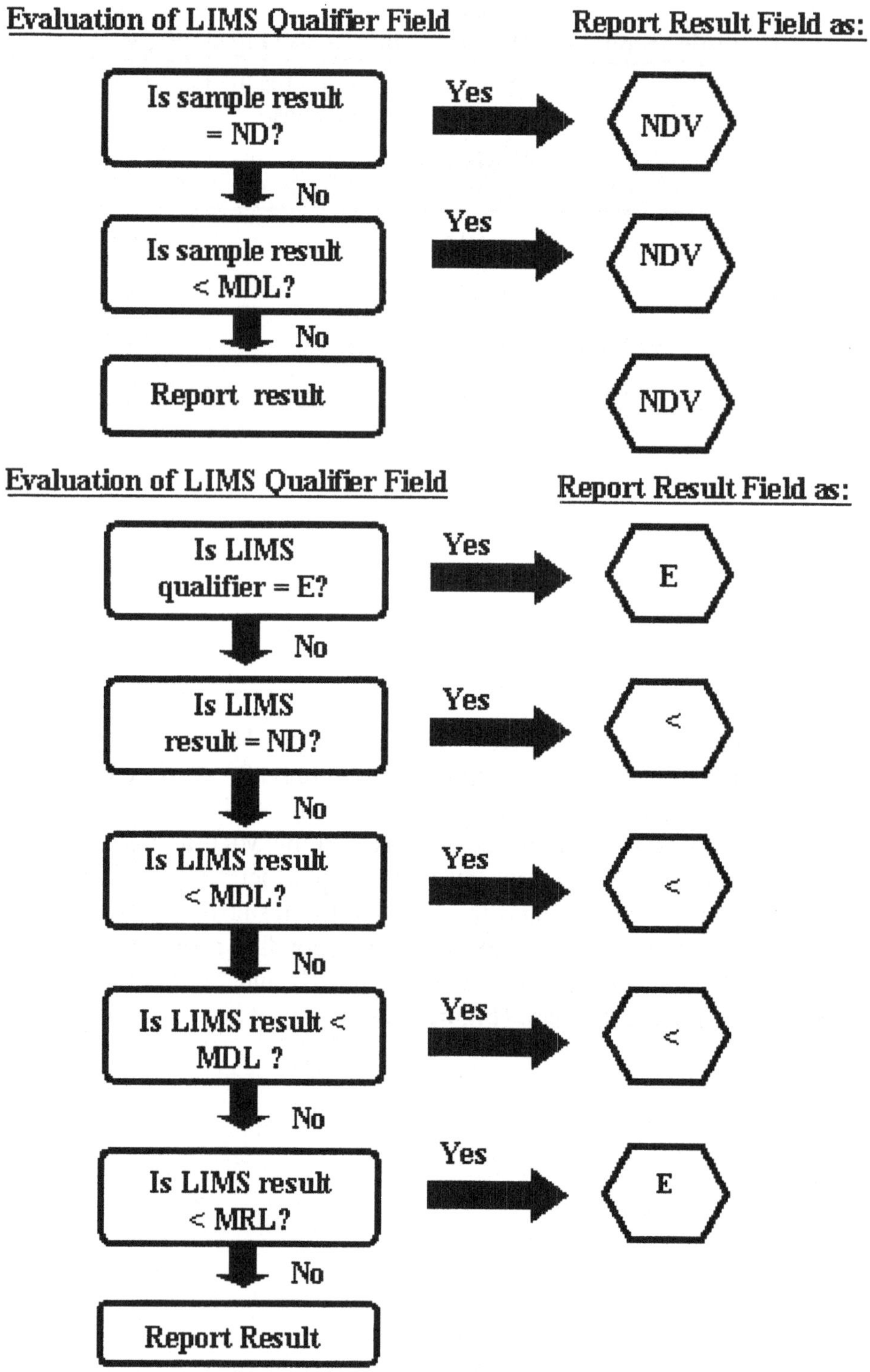

Figure 4.2 Qualification Strategy for Laboratory Information Management System (LIMS) Query and Transfer to USGS

The evaluation of the environmental samples against the MDL and MRL was performed by an Access query and created two separate linked tables. These tables contained the environmental sample information and the LIMS sample number (which was unique) and related the results. The tables were reviewed to verify that the data evaluation described in Figure 4.2 was performed accurately. The two tables were then emailed to USGS. The data transfers to USGS were performed periodically to provide updates of sample analysis.

FIELD QUALITY CONTROL/DATA ACCEPTANCE CRITERIA

Random Survey

Field quality-control (QC) data collected in conjunction with the Random Survey were reviewed and the results are presented in Appendix G. FRB and TRB samples were used to measure any systematic or random contamination from the environment around the sample site and during the sample shipment. A bottle containing VOC-free water was provided to the CWSs participating in the random source-water survey with each sample kit. Instructions were given to fill the empty VOC sample vials labeled "**Field Blank**" at each sample site with the organic free water provided in the sample kit at the same time the source-water sample was collected. TRB vials were prepared at the laboratory by filling with organic free water. TRBs were then placed in each sample kit, traveled along with the samples, and were not opened until the time of analysis at the laboratory. TRB samples were never opened and were intended to measure any contamination associated with samples during shipment.

In addition to data for 956 source-water samples, the MWDSC Water Quality Laboratory reported VOC analyses for 290 reagent-water FRBs and 112 TRBs. Although field blank vials and VOC-free reagent water were provided to all CWSs, not all FRBs were received or analyzed. A small number (21) of CWSs omitted processing and returning the FRB with their sample vials (or their FRBs were broken in transit), and the FRBs and TRBs generally were not analyzed unless VOCs were detected in the companion source-water samples. TRBs were not included in the sample kits until week 16 of the Random Survey.

Review of Field Quality-Control Data

The QC data indicate that some samples may have been contaminated because 93 percent of the FRBs and 92 percent of the TRBs analyzed contained one or more VOCs at concentrations equal to or greater than the MDL. Forty-three of the 56 VOCs detected in source-water samples were reported in one or more of the 290 FRBs at concentrations equal to or greater than the MDL; and 20 of these compounds were detected in one or more of the 112 TRBs (Appendix G - Table G.23). Toluene and methylene chloride were the most frequently detected VOCs, reported at concentrations equal to or above their MDLs in nearly three-quarters of the FRB and TRBs, but nine additional VOCs, including MTBE, were reported in more than 10 percent of the FRBs and TRBs (Appendix G - Table G.23). The contamination was largely low-level, however, and most (80 percent) VOC concentrations in the field and trip reagent blanks were less than the MRL (equal to 0.2 µg/L for all VOCs except for methyl ethyl ketone, which has an MRL of 2.0 µg/L). Still, 25 of the VOCs detected in FRBs and 8 of the VOCs detected in TRBs were measured in some samples at concentrations equal to or above the MRL.

The extensive low-level contamination evident in the field QC samples prevented reporting the occurrence of VOCs in source water at concentrations below the MRL; however, because contamination was measured in 108 FRBs (37 percent) and 10 TRBs (8.9 percent) at concentrations at or above the MRL, the field QC data were analyzed further to determine the level of uncertainty associated with VOC detection frequencies in source water reported at the MRL. The procedures used to evaluate and quantify the extent of external sample contamination and the implications for the source-water findings are described in Appendix G.

Source-Water Data Acceptance Criteria

The review of field QC and source-water samples suggested that random contamination of some source-water samples may have occurred for some VOC analytes even at concentrations censored at the MRL. The strategy ultimately selected to calculate the frequency of detection of each analyte in the Random Survey was to exclude detections for source-water samples when evidence of possible random contamination for that analyte existed (condition E2 in Tables G.32 and G.33). As such, the occurrence of VOCs in source water reported by the Random Survey has been adjusted downward for the extent of possible random contamination determined from the field QC samples collected as part of this survey.

Collectively, laboratory and field quality-control data and the procedure described in Appendix G used to calculate VOC detection frequencies, provided a high assurance of the validity of reported detections. To be included as a valid detection in the Random Survey, a compound tentatively identified from the analysis of a source-water sample must have passed all three of the following criteria:

1. The compound was positively identified via its mass fragmentation pattern;
2. The compound's detection was not due to laboratory contamination; and
3. The compound's detection was not due to field contamination.

Focused Survey

A review of the field QC data collected in the Focused Survey, as well as the source-water acceptance criteria for the Focused Survey has been conducted with parallel protocols as those of the Random Survey. FRB and TRB samples were collected in both surveys, however, in addition FEBs for surface water samples and field duplicate samples were collected in the Focused Survey. Each of these sample types is described and the results are summarized in Appendix G. In addition to this general QC review, the screening procedure used to accept or reject source-water data due to potential field contamination and justification of using source-water data below the MRL are presented here. The screening procedure, which compares FRB concentrations to source-water sample concentrations, is the same procedure used in the Random Survey to accept or reject source-water data at both the MDL and the MRL.

Review of Field Quality-Control Data

As in the Random Survey, FRB samples were used to measure contamination from the environment around the sample site and during sample shipment. In the Focused Survey, a bottle containing VOC-free water was provided to the USGS field personnel with each sample kit. FRB samples were prepared by directly filling VOC 40 mL vials labeled "Field Blank" at each sample site with VOC-free water immediately prior or following the collection of the source-water sample. USGS field personnel preserved the FRB with a 50% hydrochloric acid solution to a pH of 2. TRBs, however, were prepared at MWDSC Water Quality Laboratory by filling vials with VOC-free water. TRBs were placed in each sample kit and traveled along with the source-water samples back to the MWDSC Water Quality laboratory.

FEBs were intended to represent all contamination associated with the surface-water sampling process, sampling equipment, processing, transportation, and laboratory handling/analysis. FEB samples were collected by passing VOC-free water through the USGS VOC hand-sampler and filling the vials accordingly. FEB collection procedures for the VOC hand-sampler are described by Shelton (1997). Because ground-water samples were collected from permanently installed pumps, FEBs were collected only for surface-water samples. Field duplicate source-water samples were collected at about 15 percent of the surface- and ground-water sites immediately after an environmental sample was collected to characterize variability associated with sampling, processing, transportation, and analysis.

A total of 744 field QC samples were collected as part of the Focused Survey (Table 4.2). FRBs were collected in association with each source-water samples. However, FRB samples were collected only in association with about 15 percent of source-water samples during the first three months of sampling. One hundred percent of source-water samples were collected in association with a TRB. FEB samples and duplicate source-water samples were prepared in association with 15 percent of source-water samples collected. Finally, a total of 69 field duplicate samples were collected and analyzed. Ground-water field duplicates were collected sequentially, and surface-water field duplicates were collected concurrently. The results of the QC samples are presented in Appendix G.

Statistical Comparison of Quality Assurance/Quality Control Blanks

The extensive low concentration (>MDL) contamination evident in the field quality-control data required further analysis to determine the level of uncertainty associated with VOC detection frequencies in source water reported at or above the MDL. Similarities between detected concentrations in different blank types may indicate a common source of contamination, whereas, differences may indicate more of random source. The statistical analysis included comparisons of FRBs to TRBs and FEBs (Appendix G).

Table 4.2
Summary of the frequency and number of quality-control samples collected during the Focused Survey

Quality-control sample type	Frequency of quality-control sample collection (percent)	Number of quality-control samples analyzed
Trip-reagent blank	100	304
Field-reagent blank*	100	321
Field-equipment blank†	15	50
Field Duplicate source-water samples	15	69
Total	NA	744

* Sample frequency was 15percent during the first three months of sample collection and 100percent thereafter.
† for surface-water sites only
NA - not applicable

Source-Water Data Acceptance Criteria and Use of Detections Above the MDL

The Focused Survey had two main purposes: (1) to determine the temporal variability and (2) characterize sampling sites that had frequent occurrence of MTBE and other VOCs in sources of drinking water. In order to accomplish these two objectives, data analyses involved the use of concentration values to determine temporal trends as well as a presence/absence approach for site characterizations. The strength of these types of comparisons relies on the amount of data (detections) available for data analyses. That is, if a compound is determined to be present, it may be used in statistical tests to determine if the VOC is associated with some environmental risk factor. However, if the concentration of a compound detected below the MRL is censored at the MRL, low-level temporal trends and explanatory factors could be missed. It is important to note that data below the MRL can only be used in these analyses if the QC screening process supports those activities.

VOC results below the MRL but above the MDL were used in the data analysis for the Focused Survey as opposed to the Random Survey where all data was censored at the MRL. The use of VOC detections below the MRL (those detections between the MDL and MRL) obtained in the Focused Survey is justified because the amount of contamination was less in comparison to the Random Survey and the QC screening process removed potentially contaminated data from the data set. The reasons for the improved quality of source-water and QC data obtained from the Focused Survey are unknown. However, two possibilities include: (1) the collection of source-water samples in environments that were less contaminated in comparison to the Random Survey and/or (2) the use of a common and verified sample-collection protocol by trained USGS sampling personnel. Thus, the evaluation of data for the Focused Survey deviates from the strategy used in the Random Survey as it relates to acceptance/rejection of source-water data below the MRL. The process of screening source-water data for field contamination as well as supporting evidence of reduced contamination in the Focused Survey data set is presented in Appendix G.

DATABASE VERIFICATION

Random Survey

Analytical Data – MWDSC – MTBE/VOCs

The data verification strategy involved the complete path taken by the data (from the analysis instrument to the results database). First, data were electronically transferred from the GCMS data system to MWDSC's LIMS. Data in the LIMS underwent editing in accordance with the strategy for determining non detects and "estimated" values based on the MRL and MDL for each compound (see section "Data Quality Objectives – Data Validation"). Results were then queried from the LIMS using Microsoft Access, and the MS Access data base file was transported as an email attachment to USGS.

Verification procedures were required at all steps to ensure that the data transferred accurately and without errors. Hand-entered or edited data were always verified by a second analyst to eliminate transcription errors. Procedures for verification of the data along its electronic path from one system to another were also in place to ensure the integrity of the data.

The GCMS system transferred all values greater than the MDL to LIMS. These results were electronically imported into the MWDSC Water Quality Laboratory LIMS. There were three reporting fields in LIMS for each compound: 1)"qualifier", 2) text, and 3)numeric "result". The default for the text field was set to "not detected" (ND), if there was nothing in the numeric field. On occasion, the data were not transferred electronically to LIMS and hand entry into LIMS was required. The electronic transfer to LIMS was also checked for accuracy. The data in LIMS were reviewed to verify that the transfer was accurate and whether each analyte detected was a good spectral match. Any changes were edited and recorded in LIMS. After the data were reviewed in LIMS, the status for each sample number was changed to approved.

MWDSC provided a final analytical database to USGS that contained two separate linked tables. These tables contained the sample identification information and the analytical results linked by the LIMS sample number (which was unique). The tables were reviewed to verify that the data evaluation described in the Data Quality Objective section (Figure 4.2) was performed accurately. The two tables were then emailed as attachments to USGS. No further manipulation or verification was performed by USGS on the analytical database.

Ancillary Data

Ancillary information about the participating CWSs and their sources was collected to allow additional interpretation of the water-quality data. A written questionnaire was included in the sampling supplies sent to participating CWSs by MWDSC. Completed questionnaires were returned to MWDSC, where photocopies were made and forwarded to USGS. USGS created a database to house the ancillary data using Microsoft Access. Data were keyed into the Access database by USGS personnel using electronic input forms modeled after the paper questionnaire.

Data from all 941 questionnaires received from MWDSC or directly from CWS contacts were entered into the MS Access database maintained by USGS. These included data from 576 ground water and 365 surface water questionnaires. Questionnaires for all 940 CWSs in the U.S. were obtained, as well as one questionnaire for a CWS in Puerto Rico. Thirteen CWSs in Puerto Rico did not return a questionnaire. No effort was made to obtain these missing questionnaires because of communication and personnel constraints, and because additional GIS-derived ancillary data were not available for Puerto Rico.

Source-water locations (latitude/longitude) were obtained and verified for all 940 CWS sources in the U.S. through phone calls to each of the CWSs made by USGS. At the same time, an effort was made to obtain ancillary data missing from the questionnaires. This effort was 100% successful in obtaining source-water locations, and was successful in obtaining most of the missing ancillary data. For each site, the USGS caller assigned an accuracy category (based on confidence in the site determination) with an estimated distance that the determined site may differ from the true location.

Latitude/Longitude accuracy code

Accuracy categories are numerically assigned designations that represent the approximate number of feet that a source-water location (assigned by latitude-longitude) can be determined with confidence. The latitude-longitude for each source was established by contacting CWS representatives via the telephone. During an interchange, the Delorme Street Atlas USA Computer Program was used to geographically locate the source location. The telephone contact gave verbal instructions, generally by street names and directions, for the location of the source-water sample location. The caller tracked these directions via the computer program. The "readout" latitude-longitude and the graphic distance scale that appeared on the computer screen were used to assign accuracy categories.

Because distances on the Delorme scale bar are in decimal miles only, value ranges for the accuracy categories lack a numerically systematic appearance when converted to feet. Also, in the degree-minute-second (DMS) latitude-longitude description, the second (S) is rounded off to the nearest tenth, which represents approximately ten feet geographically. It was determined that the degree of accuracy could not be determined to the nearest one hundredth of a second (representing a one-foot distance). Latitude-longitudes for source water samples are thereby described in DMS to the nearest one-tenth second. Accuracy categories and associated distances are shown in Table 4.3.

A subsequent effort involved verifying the information in the MS Access database using the paper questionnaires. Select fields deemed to have the most importance (such as source location) by the USGS scientists involved in interpretation of the data were compared between the paper and electronic files, and any discrepancies were corrected. The verification step was performed on all 941 questionnaires received.

Table 4.3
Accuracy categories and distances for latitude/longitude determination

Accuracy Category	Distance in feet	Number of Source Waters GW	SW
1	10	132	66
2	11 – 53	250	197
3	54 – 264	141	90
4	> 265	52	12
Total		575	365

gw – ground water source   sw – surface water sources

A further verification of source-water locations was performed using GIS. An overlay of source-water locations was made on a state layer. The state code contained within the PWSID for each CWS (for example, "CA" in CA0092543) was compared to the code for the state in which the site was plotted. Any discrepancies were checked within the paper questionnaire files and corrections were made wherever necessary. Additionally, any site that plotted outside of the U.S. boundary (for example, in Lake Michigan) was checked to make sure that the designation was correct (for example, an intake three miles out in the Lake).

Focused Survey

Analytical Data – MWDSC – MTBE/VOCs

See section for Random Survey above.

Analytical Data – OHSU – MTBE By-products

The Oregon Health & Science University (OHSU) provided 32 separate data files with analytical results to USGS as email attachments. The provided files were in Microsoft Excel format and listed site names as columns and compound names as rows. Reformatting of each file was done by USGS to list compound names as columns and site names as rows. The reformatting was performed to allow ease of linkage with the MWDSC data, which was in the compound name-column site name-row format, and to allow easy linkage to ancillary data, which was arranged in rows by site name. All reformatted data files were returned to OHSU as email attachments and were verified by OHSU staff to ensure that no errors were created in the reformatting process. No errors were found. A final comprehensive Excel file was then imported into Microsoft Access for querying and relating to other data sets.

Ancillary Data

A written questionnaire was provided to the USGS personnel responsible for sampling source waters in the Focused Survey. USGS sampling personnel were asked to provide the questionnaire to the CWS contacts, who then filled out the questionnaires and returned them by mail to the USGS office in Rapid City. Additionally, USGS sampling personnel were asked to mark off the location of the source water site on a 1:24,000 USGS topographic map and return it to the USGS Rapid City office.

Data from 120 questionnaires were entered into an MS Access database created by USGS to house the ancillary data. These included data for 78 ground water and 58 surface water sources. One questionnaire was never returned for a CWS with two ground-water sources. However, source-water locations (latitude/longitude) were obtained for all sources in the Focused Survey. These locations were plotted using DeLorme Street Atlas USA so that a consistent read-out in the same horizontal datum could be obtained for all sites. Site verification calls to CWS contacts were not made for the Focused Survey sites because actual site visits were made by USGS sampling personnel. Accuracy categories were not assigned to the sites because all sites were assumed to be equally accurate to within 10 feet.

A subsequent effort involved verifying the information in the MS Access database using the paper questionnaires. All fields, including source-water locations, were compared between the paper and electronic files, and any discrepancies were corrected. The verification step was performed on all 120 questionnaires received.

CHAPTER 5
FINDINGS

RANDOM SURVEY

Implementation of the Random Survey

The Random Survey was designed to ensure an unbiased distribution of CWSs by state, type of source water (ground water or surface water), and population served (Ivahnenko, Grady, and Delzer 2001). A statistical sampling of a subset of CWSs would allow information on the frequency of occurrence and concentration of MTBE and other VOCs in drinking-water sources for randomly selected CWSs to represent, in aggregate, the overall population of CWSs.

Data on the distribution of the nearly 47,000 active, self-supplied CWSs that serve more than 250 million people (Table 5.1) were obtained from the USEPA's Safe Drinking Water Information System (SDWIS) on November 5, 1998 (U.S. Environmental Protection Agency 1998b). By design, the Random Survey would distribute 1,000 samples among 10 source-size categories (5 for ground water, 5 for surface water) stratified to reflect the national distribution of the self-supplied systems and the total number of people served by CWSs within each category (Ivahnenko, Grady, and Delzer 2001). Because more than 60 percent of CWSs are very small (serve 25 to 500 people) and are supplied by ground-water sources, but 45 percent of the population served by CWSs are customers of very large (more than 50,000 people served) surface-water supplied systems, population served and source-water type were weighted equally in the design of the Random Survey. Consequently, the distribution of the 1,000 CWSs in the Random Survey included 613 mostly very small and small ground-water-supplied systems, and 387 predominantly large and very large surface-water-supplied systems.

The Random Survey was conducted over a 78-week period that began on May 3, 1999 and ended on October 23, 2000. Selection of participating CWSs was made from randomized lists of the 46,960 active, self-supplied water utilities obtained from the SDWIS database on November 5, 1998, until the requisite number of systems was obtained for each source-size category in each state, Native American Lands, and Puerto Rico. Subsequently, sample-collection kits were distributed to the selected CWSs.

A total of 954 source-water samples were submitted to the MWDSC laboratory by the CWSs randomly selected for participation in the survey. With participation by 95 percent of the CWSs included in the design, the sample size achieved was sufficiently close to the total 1,000 planned systems for the Random Survey to accomplish the designed distribution and to allow valid statistical analysis.

Table 5.1
Number of self-supplied community water systems and number of people served, by type of source water and size of system, November 5, 1998*

	Ground water				Surface water			
CWS size category	Number of systems	Percent of all systems	Number of People served	Percent of total population served	Number of systems	Percent of all systems	Number of People served	Percent of total population served
Very small	28,324	60.3	4,625,130	1.84	1,228	2.64	616,012	0.24
Small	9,775	20.8	14,178,037	5.63	1,562	3.33	5,739,217	2.28
Medium	2,399	5.11	14,219,831	5.65	971	2.07	11,045,463	4.39
Large	1,194	2.54	25,342,137	10.1	928	1.98	36,525,585	14.5
Very large	182	0.39	25,696,338	10.2	397	0.85	113,671,630	45.2
Total	41,874	89.1	84,061,473	33.4	5,086	10.9	167,597,907	66.6

* - Data from U.S. Environmental Protection Agency's Safe Drinking Water Information System (URL http://www.epa.gov/enviro/html/sdwis/)

CWS - community water system

CWS size categories: very small, serving less than 500 people; small, serving 501 to 3,300 people; medium, serving 3,301 to 10,000 people; large, serving 10,001 to 50,000 people; very large, serving more than 50,000 people

The distribution of the 954 CWSs (that participated) by source-water type and system size is shown in Table 5.2. The percentage of ground-water sources sampled for each of the five size categories ranged from 92 to 100 percent of the number planned in the design, and the percentage of the surface-water sources sampled ranged from 86 to 99 percent of the number planned.

Table 5.2
Number of community water systems planned and sampled for the Random Survey, by type of source water and size of system.

	Type of source water and number of systems planned			Type of source water And number of Systems sampled			Type of source water and percent of planned systems sampled		
CWS size category	Ground water	Surface water	Total	Ground water	Surface water	Total	Ground water	Surface water	Total
Very small	311	14	325	292	12	304	94	86	94
Small	132	28	160	121	26	147	92	93	92
Medium	54	32	86	50	30	80	93	94	93
Large	63	83	146	63	79	142	100	95	97
Very large	53	230	283	53	228	281	100	99	99
Total	613	387	1,000	579	375	954	94	97	95

CWS - community water system

CWS size categories: very small, serving fewer than 500 people; small, serving 501 to 3,300 people; medium, serving 3,301 to 10,000 people; large, serving 10,001 to 50,000 people; very large, serving more than 50,000 people

Geographic Distribution of Participating Community Water Systems

The geographic distribution of systems sampled was within 90 percent of the target proportion for 45 of the 52 states or other geographic entities (Table 5.3). The only substantial difference in the achieved distribution of CWSs from the design was the selection of only 14 of 22 planned systems in U.S. Territories, and that all 14 of these were in Puerto Rico. The logistical difficulties of locating and contacting potential participating CWSs in other U.S. Territories required this modification in design. Overall, the achieved distribution has a slight deficit in the number of smaller systems for both surface- and ground-water-supplied CWSs. Generally, it was considerably more difficult to obtain participants from the smallest size categories, and the total population of very small surface-water supplied systems is limited. For some states, failure by the one or two systems in these categories to participate in the survey provided no alternative selections.

Table 5.3
Number of community water systems sampled and planned for the Random Survey, by source-size category and State.

State or other entity	Number of CWSs sampled / number of CWSs planned											Percentage of planned
	Source-size category											
	GW-VSM	GW-SM	GW-MED	GW-LRG	GW-VLRG	SW-VSM	SW-SM	SW-MED	SW-LRG	SW-VLRG	Total	
AK	4/4	1/1	0	0	0	1/1	0	0	0	1/1	7/7	100
AL	0/0	2/2	2/2	1/1	0	0	0	1/1	3/3	5/5	14/14	100
AR	2/2	2/2	1/1	0	0	0	0/1	1/1	1/1	1/1	8/9	89
AZ	5/5	2/2	1/1	2/2	1/1	0	0	0	0	6/6	17/17	100
CA	19/20	5/5	4/3	7/7	11/11	2/2	2/2	2/2	5/5	30/30	87/87	100
CO	4/5	1/1	0	0	0	1/1	1/1	0/1	2/2	6/6	15/17	88
CT	5/5	1/1	0	0	0	0	0	0	2/2	5/5	13/13	100
DE	2/2	1/1	0	0	0	0	0	0	0	1/1	4/4	100
FL	13/13	5/6	3/3	7/7	17/17	0	0	0	1/1	3/3	49/50	98
GA	11/12	2/3	1/1	1/1	1/1	0	1/1	1/1	3/3	8/8	29/31	94
HI	0	1/1	0	0	1/1	0	0	0	0	0	2/2	100
IA	7/7	4/4	1/1	1/1	1/1	0	0	0	0	2/2	16/16	100
ID	5/5	1/1	1/1	1/1	0	1/1	0	0	0	1/1	10/10	100
IL	7/7	4/5	2/2	2/2	1/1	0	1/1	1/1	3/3	8/8	29/30	97
IN	4/4	4/4	2/2	2/2	1/1	0	0	1/1	1/1	4/4	19/19	100
KS	4/4	3/3	0	1/1	0	0	1/1	1/1	1/1	3/3	14/14	100
KY	1/1	0	0	0	0	0	1/1	2/2	4/5	2/2	10/11	91
LA	8/8	5/5	1/3	2/2	½	0	0	1/1	1/1	5/5	24/27	89
MA	2/2	0/1	2/2	3/3	1/1	0	0	1/1	4/4	8/8	21/22	95
MD	3/4	1/1	1/1	1/1	0	0	0	0	1/1	5/5	12/13	92
ME	2/2	1/1	0	0	0	0	1/1	0	1/1	0	5/5	100
MI	8/9	4/4	1/1	1/1	1/1	0	0	1/1	2/2	4/4	22/23	96
MN	5/6	4/4	1/1	3/3	0	0	0	0	1/1	2/2	16/17	94
MO	8/8	4/4	2/2	1/1	1/1	0	1/1	1/1	1/1	4/4	23/23	100
MS	5/5	8/9	3/3	2/2	0	0	0	0	0	0	18/19	95
MT	4/5	1/1	0	0	0	0	0	0	0	1/1	6/7	86
NC	15/16	3/3	1/1	1/1	0	0	1/1	2/2	4/4	7/7	34/35	97
ND	1/1	1/1	0	0	0	0	0	0	1/1	0	3/3	100
NE	5/5	2/2	1/1	1/1	0	0	0	0	0	1/1	10/10	100
NH	5/5	1/1	0	0	0	0	0	0	1/1	1/1	8/8	100
NJ	3/3	2/2	2/2	4/4	1/1	0	0	0	1/1	8/8	21/21	100
NM	5/5	1/1	1/1	1/1	1/1	0	0	0	0	0	9/9	100
NV	2/2	1/1	0	0	0	0	0	0	0	1/1	4/4	100
NY	12/15	3/4	1/1	3/3	4/4	1/1	2/2	2/2	4/4	16/16	48/52	92
OH	7/7	4/4	2/2	3/3	2/2	0	1/1	1/1	4/4	10/10	34/34	100
OK	3/3	1/2	1/1	1/1	0	1/1	2/2	2/2	2/2	4/4	17/18	94

(continued)

Table 5.3 (continued)

State or other entity	Number of CWSs sampled / number of CWSs planned Source-size category GW-VSM	GW-SM	GW-MED	GW-LRG	GW-VLRG	SW-VSM	SW-SM	SW-MED	SW-LRG	SW-VLRG	Total	Per-centage of planned
OR	5/6	1/1	0	1/1	0	1/1	1/1	1/1	2/2	2/2	14/15	93
PA	11/14	4/5	2/2	1/1	0	1/1	1/1	2/2	6/6	15/15	43/47	92
RI	0	0	0	0	0	0	0	0	0	1/1	1/1	100
SC	3/4	1/1	1/1	0	0	0	0	0	2/2	3/3	10/11	91
SD	2/2	1/1	0	0	0	0	0	0	0	1/1	4/4	100
TN	1/1	1/1	1/1	1/1	1/1	0	1/1	2/2	4/4	4/4	16/16	100
TX	25/24	12/14	3/6	3/3	3/3	0/1	1/2	2/2	5/5	17/17	71/77	92
UT	3/3	1/1	1/1	1/1	0	0	0	0	0	4/4	10/10	100
VA	11/11	2/2	0	0	0	0	1/1	1/1	2/2	6/6	23/23	100
VT	2/3	1/1	0	0	0	0	1/1	0	0	0	4/5	80
WA	15/17	4/4	1/1	2/2	1/1	1/1	1/1	0	1/1	2/2	28/30	93
WI	7/7	4/4	2/2	1/1	1/1	0	0	0	1/1	3/3	19/19	100
WV	2/2	0/1	0	0	0	0	2/2	1/1	1/1	1/1	7/8	88
WY	2/2	0	0	0	0	0	0	0	0	0	2/2	100
NA	6/6	2/2	0	0	0	1/1	0	0	0	0	9/9	100
PR	1/2	1/1	1/1	1/1	0	1/2	2/2	0/1	1/4	6/8	14/22	64
Total	292/ 311	121/ 132	50/ 54	63/ 63	53/ 53	12/ 14	26/ 28	30/ 32	79/ 83	228/ 230	954/ 1,000	95
Percen-tage of planned	94	92	93	100	100	86	93	94	95.	99	95	

GW - ground water; SW - surface water; VSM - very small; SM - small; MED - medium; LRG- large; VLRG - very large; NA- Native American Lands; PR- Puerto Rico

The percentage of planned systems sampled in each state ranged from 67 to 100 percent for ground-water supplied systems and from 59 to 100 percent for surface-water supplied systems. A small deficit in participating ground-water systems occurred in almost half the states, mostly for the very small and small systems. The very small systems typically had part-time operators and management, and often there were difficulties in making contact with or obtaining the samples from very small ground-water supplied systems.

The geographic distribution of the 579 ground-water sources and 375 surface-water sources in the 50 states, Native American lands, and Puerto Rico sampled for the Random Survey are shown in Figure 5.1. Nearly all (98 percent) of the ground-water sources were wells (2 percent were springs). Surface-water sources included 204 lakes and (or) reservoirs (54 percent), and 171 rivers, streams, aqueducts, or canals (46 percent). Fifty-five percent of the

Figure 5.1 Distribution of drinking-water sources for community water systems in the United States and Puerto Rico sampled for the Random Survey.

drinking-water sources sampled in the Random Survey are east of the Mississippi River. Proportionally, more CWSs in the east have surface-water sources than those in the west, where ground-water sources make up two-thirds of source waters sampled. The distribution of source waters also reflects regional physiographic and hydrogeologic characteristics. Ground-water sources are predominant in much of the southeastern Atlantic and Gulf of Mexico coastal plain in Georgia, Florida, Mississippi, and Louisiana, and in the High Plains that are underlain by productive aquifers. Surface-water sources are more dominant in the humid Appalachian and Ozark Mountain areas and along the shores of the Great Lakes.

Temporal Distribution of Source-Water Samples

Sample collection for the Random Survey was designed to provide an even distribution of the number of samples collected in any particular month or season. The temporal distribution of the 1,000 source-water samples planned for in the design of the Random Survey was to spread the sample collection over a 64-week period in such a way as to preclude seasonal bias in the data (Ivahnenko, Grady, and Delzer 2001). To achieve an equal number of samples on a monthly basis for a calendar year, about 83 samples (or 8 percent of the samples) should have been collected each month. Because the sampling was planned to extend beyond a calendar year, only about one-half as many samples were scheduled to be collected each week during the initial and final 12 weeks of the survey (collectively) as were scheduled during the intervening 40 weeks (Figure 5.2). The actual sampling period, however, extended for 78 weeks (from May 3, 1999 through October 23, 2000) as logistical considerations (such as holiday work schedules) and the need to reselect and (or) reschedule some participating CWSs caused deviation from the designed sample frequency. Because of the difficulties experienced in locating the requisite number of small CWSs to participate in the Random Survey, the sample period (Figure 5.2) was extended to allow for participation by as many CWSs as could be arranged. The modifications in the sampling schedule did skew the temporal pattern of the samples from the even distribution planned. The percentage of all samples collected during any particular month of the year exceeded the planned 8 percent per month during the summer months (Figure 5.3). In particular, the number of samples submitted during August was almost 150 percent of the target, whereas a deficit occurred during the fall and early winter months, especially November and December. The geographic distribution of drinking-water sources sampled in any month was random, however, and it is considered unlikely that the occurrence of MTBE or other VOCs determined during the Random Survey was affected by the temporal variations in sample allocation.

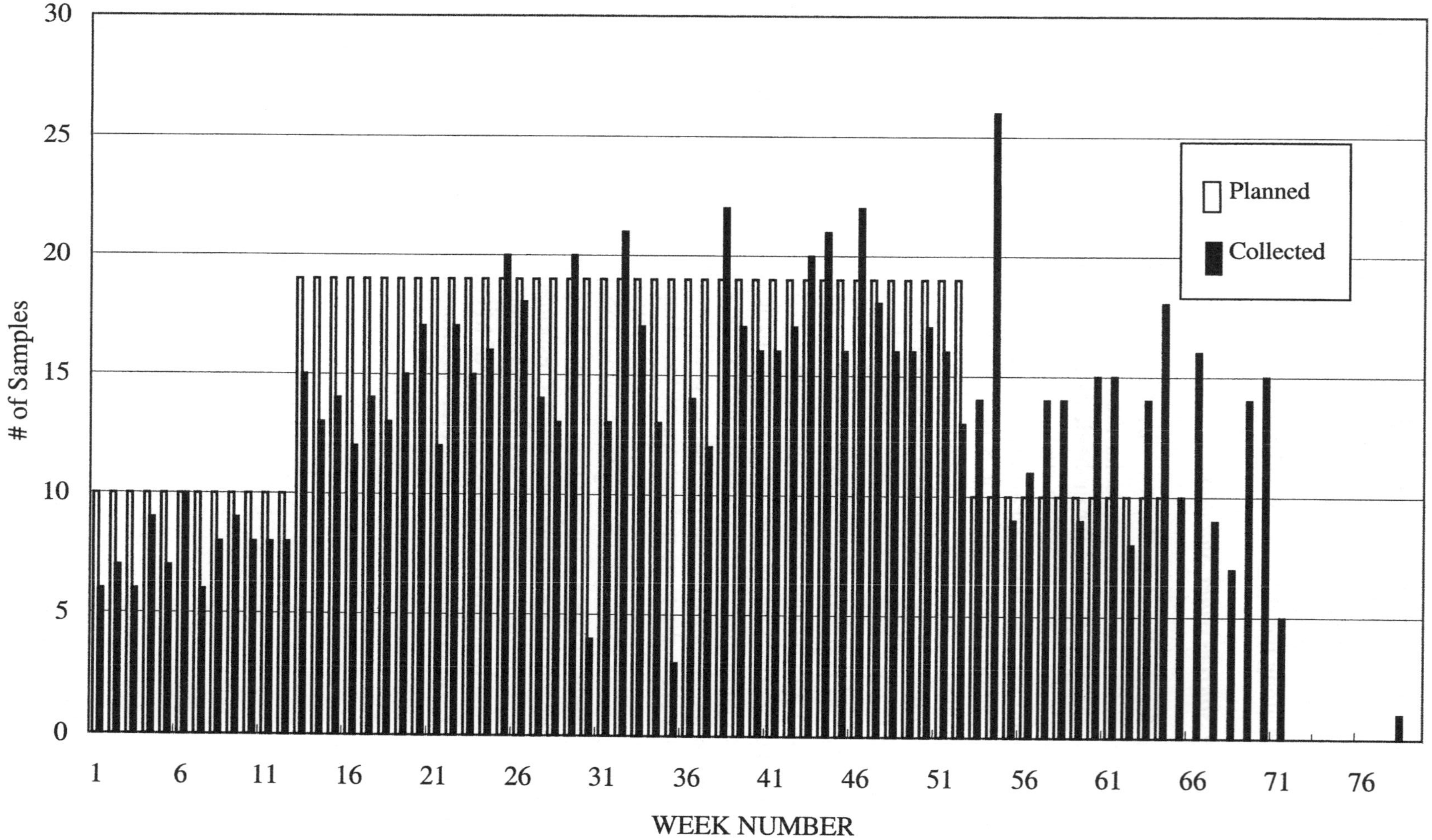

Figure 5.2 Number of source-water samples planned and the number of samples collected per week for the Random Survey

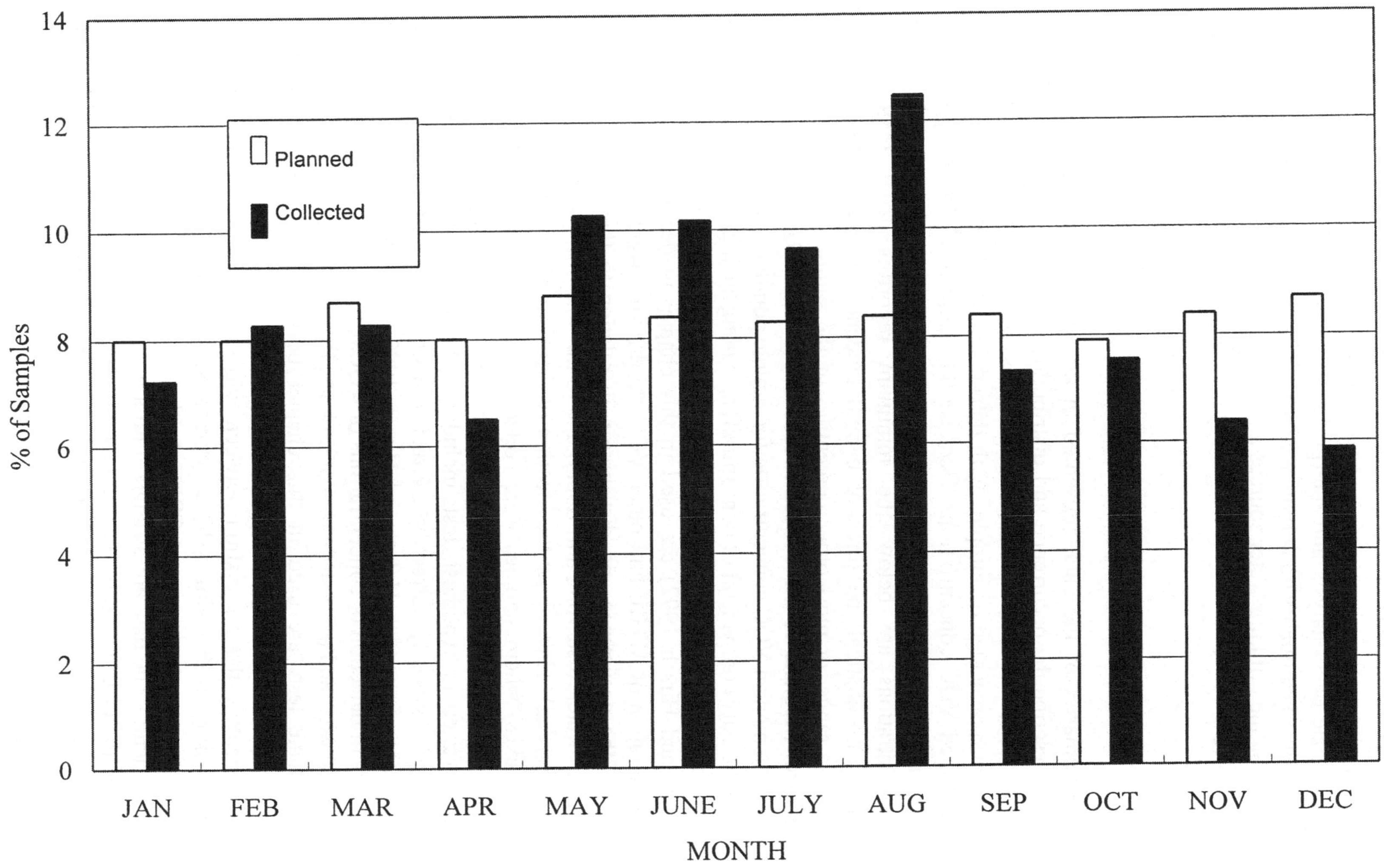

Figure 5.3 Percentage of source-water samples planned and the percentage of samples collected per month for the Random Survey

Data Collection and Analysis

The findings of the Random Survey are based on information obtained and reviewed as part of four data collection and analysis activities described in this report: (1) collection and laboratory chemical analysis of source-water and companion field, quality-control samples, (2) application of appropriate data analysis and interpretation methods, (3) review of the quality-control data to assure that data-quality objectives were met, and (4) collection and compilation of ancillary information on the drinking-water sources and other geospatial data that may contribute to a better understanding of the occurrence and distribution of MTBE and other VOCs in source water.

Analysis of Data

Statistical summaries of the data, presented in tabular and graphical formats in this chapter, are used to describe the occurrence and distribution of MTBE and VOCs in drinking-water sources. Descriptive statistics include number of samples, number of detections, the frequency of detection (percentage of samples with detections), detectable concentration range, and medians of detected VOC concentrations. Because the data for all VOCs reported by the MWDSC laboratory are highly censored, that is, 50 percent or more (and commonly 90 percent or more) of concentrations are below the minimum reporting level (MRL), median concentrations or other statistical measures of central tendency (mean) or spread (standard deviation, interquartile range, or most other percentiles of the sample-population distribution) cannot be determined. Medians of the concentrations above the MRL, although a positively biased indication of the median VOC concentration for sample populations, are provided because this statistic affords the only comparison of concentrations among such highly censored analytes.

Descriptive and nonparametric statistics, histograms, cumulative frequency plots, and scatter plots (Helsel and Hirsch 1992) are used in this chapter to describe the frequency of detection and concentration of MTBE and other VOCs in drinking-water sources. The data are summarized to show the occurrence and distribution in relation to the type of source water (ground water, surface water, reservoirs and rivers) and to the size of the CWSs (five categories based on population served). Maps show the location of drinking-water sources (wells, springs, and surface-water intakes) where samples were collected for analyses of MTBE and selected other VOCs, and the location of drinking-water sources where these compounds were detected. Where appropriate, a variety of hypothesis tests including contingency-table tests (Pearson's chi-square test with Yates' continuity correction), Kruskal-Wallis tests, and the Wilcoxon signed rank test (Iman and Conover 1983, Helsel and Hirsch 1992), were used to test for relations between the frequency of detection or concentrations of VOCs and anthropogenic factors such as land use, population density, and the distribution of high MTBE-use areas.

Contingency-table tests are used in this chapter because the highly censored VOC analytical data largely precludes analysis by hypothesis tests that evaluate continuous variables. Contingency tables measure the association between two discrete, categorical variables. For example, is the probability of detecting a VOC (compared to the probability of nondetection) related to a type of source water, the size of CWSs, land use, or the presence or absence of some other anthropogenic factor? The data are arranged into a matrix of rows and columns—with no natural ordering—and the distribution of data among the categories is tested to determine if the row classification is independent of the column using the chi-square distribution (Helsel and

Hirsch 1992). As with other statistical tests used in this report, the results of the contingency-table tests are expressed by the "p-value" or the significance level attained by the data; for this report, the null hypothesis is rejected and the two variables are determined to be significantly related at p-values less than 0.05 (95-percent confidence level). When the variables are found to be dependent or related, however, it is not necessarily implied that one variable causes the observed response in the second variable.

Collection of Ancillary Information

In addition to source-water and quality-control samples, ancillary information was collected from the participating CWSs and other sources. This information was needed for statistical analysis and to determine possible relations between the occurrence of MTBE or other VOCs in the source-water samples and anthropogenic factors. Ancillary information included location (latitude and longitude) of the drinking-water sources sampled (well or intakes); actual population served by the CWS; source-water characteristics (for example, well depth, yield, aquifer type, surface-water type and size, intake specifics, and any previous water-quality problems); areal patterns of MTBE and other fuel-oxygenate use; land use; population density; and known or potential point-source locations of VOCs (Toxic Release Inventory Sites, leaking underground storage tanks, and Resource Conservation and Recovery Act and Comprehensive Environmental Response, and Compensation and Liability Act regulated sites) near the sampled drinking-water sources. This information was obtained from available data bases and (or) collected directly from participating CWSs through their responses to a mail-in questionnaire.

Information on the precise location of drinking-water sources often is missing from the SDWIS data base; therefore, latitude and longitude information was requested from each of the participating CWSs for each drinking-water source sampled (intake or well). If latitude and longitude were not available, the CWS was asked to locate the well or intake on a topographic map, from which USGS personnel determined the latitude and longitude. To confirm latitude and longitude locations (if provided) or to determine the location of the drinking-water source (if not provided), a follow-up telephone call was made during which the CWS contact was asked to verbally describe the location of the well or intake. Latitude and longitude was then determined or confirmed using DeLorme Street Atlas/Topo USA software (DeLorme, Yarmouth, ME), and it was entered into a geographical information system (GIS). The results of the analysis that use the locational information obtained for this study are included in a later section of this report, however, the locations of drinking-water sources are not releaseable under current (2002) USGS Homeland Security policy.

A short questionnaire was filled out during an initial telephone interview with CWS staff when they agreed to participate in the Random Survey. The initial questionnaire was used to verify basic information obtained from the SDWIS data base for the selected utility, obtain some additional information on the source waters to be sampled, and identify the persons and means for further contact. A more detailed written questionnaire (Random Survey Mail-In Questionnaire - Appendix B) was subsequently delivered to the CWSs with their sample kit. The mail-in questionnaire requested information about the source water, intake location, filtration and treatment, distribution area, actual population served, and the quality and quantity of water delivered by suppliers. Specific information requested on the questionnaire included well characteristics, aquifer type, vulnerability of ground-water sources, and watershed protection and watercraft-use characteristics of surface-water sources.

Participation in the mail-in questionnaire survey was remarkably high—99 percent of the CWSs participated and 941 responses to the written questionnaire were received. The responses to 70 questions posed by the questionnaire were tabulated and are presented in Appendix H – Table H.1. This information provides valuable insight into environmental conditions and operational characteristics of drinking-water sources sampled for the Random Survey. To the extent possible, this information will be extrapolated to provide support to observations and conclusions regarding the occurrence and distribution of MTBE and other VOCs in source water. To some degree, however, the utility of the ancillary information from the mail-in questionnaire is constrained by the relatively high number of responses with missing or ambiguous information for some questions (Appendix H – Table H.1).

National geospatial data on land use (Vogelmann et al. 2001), population density (Price and Clawges 1999), and point-source locations of VOCs (Vista Information Solutions 1999) were used to augment the ancillary information obtained from the mail-in questionnaire and to provide a consistent coverage for all participating CWSs. Information on MTBE or other fuel-oxygenate use was compiled from USEPA documentation (U.S. Environmental Protection Agency 1999a, 1999b) and industry surveys. This ancillary information was used to identify important natural and anthropogenic factors associated with the locations of drinking-water sources that relate to the frequency and concentration of MTBE and other VOCs observed in source water.

Occurrence and Distribution of Volatile Organic Compounds in Drinking-Water Sources

The 954 source-water samples collected from May 3, 1999 to October 23, 2000 by the participating CWSs were analyzed for VOCs by the MWDSC laboratory and are the basis for the findings of the Random Survey. The data include analyses for 66 VOCs (Chapter 3 - Table 3.2) in 579 ground-water and 375 surface-water samples. Some samples for six compounds—dibromomethane, 1,2-dichloropropane, 1,3-dichloropropane, bromochloromethane, trichloroethene, and methyl ethyl ketone— did not meet the daily calibration control checks (83 times collectively), and consequently, these analytes were not reported by the MWDSC laboratory. Also, as discussed previously in this report, a review of quality-control data identified 153 VOC detections at the MRL that may have resulted from random contamination of source-water samples (condition E2 samples in Appendix G Table G.33), and these analytical determinations were removed from the source-water database. Consequently, 22 analytes were reported in fewer than 954 source-water samples (Appendix H – Table H.2). A total of 62,728 (99.6 percent) of 62,964 potential VOC analytical determinations of the 66 target analytes are included in this analysis.

Forty-two of the 66 target VOC analytes were detected in at least one source-water sample at a concentration equal to or greater than its MRL. Eleven compounds were detected in more than 1.0 percent of source-water samples (Figure 5.4), whereas 31 VOCs were detected in less than 1.0 percent of the samples (Figure 5.5). One or more VOCs were detected in 257 (27 percent) of the 954 source-water samples at concentrations equal to or exceeding the MRL. The number of detections at or above the MRL, the frequency of detection at the MRL, and the range of concentrations of all VOCs detected in the source-water samples is tabulated by type of source water and size of CWS in Appendix H – Table H.2.

Chloroform was the most frequently detected VOC, reported at concentrations equal to or greater than 0.2 μg/L in 120 (nearly 13 percent) of the source-water samples (Figure 5.4). MTBE was the second most frequently detected VOC, reported in 81 (8.7 percent) of the source waters sampled (Figure 5.4). Other VOCs detected in 10 or more source-water samples (Figure 5.4) include three additional trihalomethanes—bromodichloromethane, chlorodibromomethane, and bromoform—several commonly used solvents or their degradation by-products—tetrachloroethene, trichloroethene, cis-1,2-dichloroethene, 1,1-dichloroethane, and 1,1,1-trichloroethane—and one other gasoline compound, toluene. All other VOCs detected in source waters (Figure 5.5) were detected rarely (in fewer than 10 source-water samples).

Although VOCs were detected in about one-quarter of the source waters sampled, VOC concentrations were low. About 95 percent of the total 530 VOC detections reported by the MWDSC laboratory were at concentrations less than 10 µg/L, and nearly two-thirds of all detections (63 percent) were less than 1.0 µg/L. However, eight source-water samples contained concentrations of 1,1-dichloroethene, tetrachloroethene, trichloroethene, and (or) vinyl chloride that would exceed USEPA's MCLs if the samples represented finished drinking water. All but one of these elevated concentrations were in samples from ground-water sources (one surface-water sample contained slightly more than 5 µg/L of tetrachloroethene and one other surface-water sample exceeded the total trihalomethane (TTHM) MCL of 80 µg/L). The cumulative distribution of the concentrations of VOCs detected in source-water samples (Figure 5.6) shows that concentrations were similar in ground water and surface water. When VOCs were detected in source-water samples, co-occurrence of several VOCs was fairly common, with multiple detections in nearly half (47 percent) of the 257 samples with VOC detections. Ground-water sources were more likely to contain multiple VOCs than surface-water sources (Figure 5.7), particularly when five or more VOCs occurred together. One ground-water sample contained 17 individual compounds and another contained 10 compounds. The percent co-occurrence among the 42 VOCs detected in source-water samples is equal to the number of times compound X also was detected among the samples that contained compound Y, multiplied by 100 (Appendix H – Table H.3). The appendix shows that 46 pairs of VOCs co-occurred at least 20 percent of the time (when 10 or more samples contained one of the paired VOCs), and that 17 pairs exhibited co-occurrence 50 percent or more of the time. Co-occurrence of VOCs in source-water samples most frequently involved detections of solvents, THMs, and gasoline compounds.

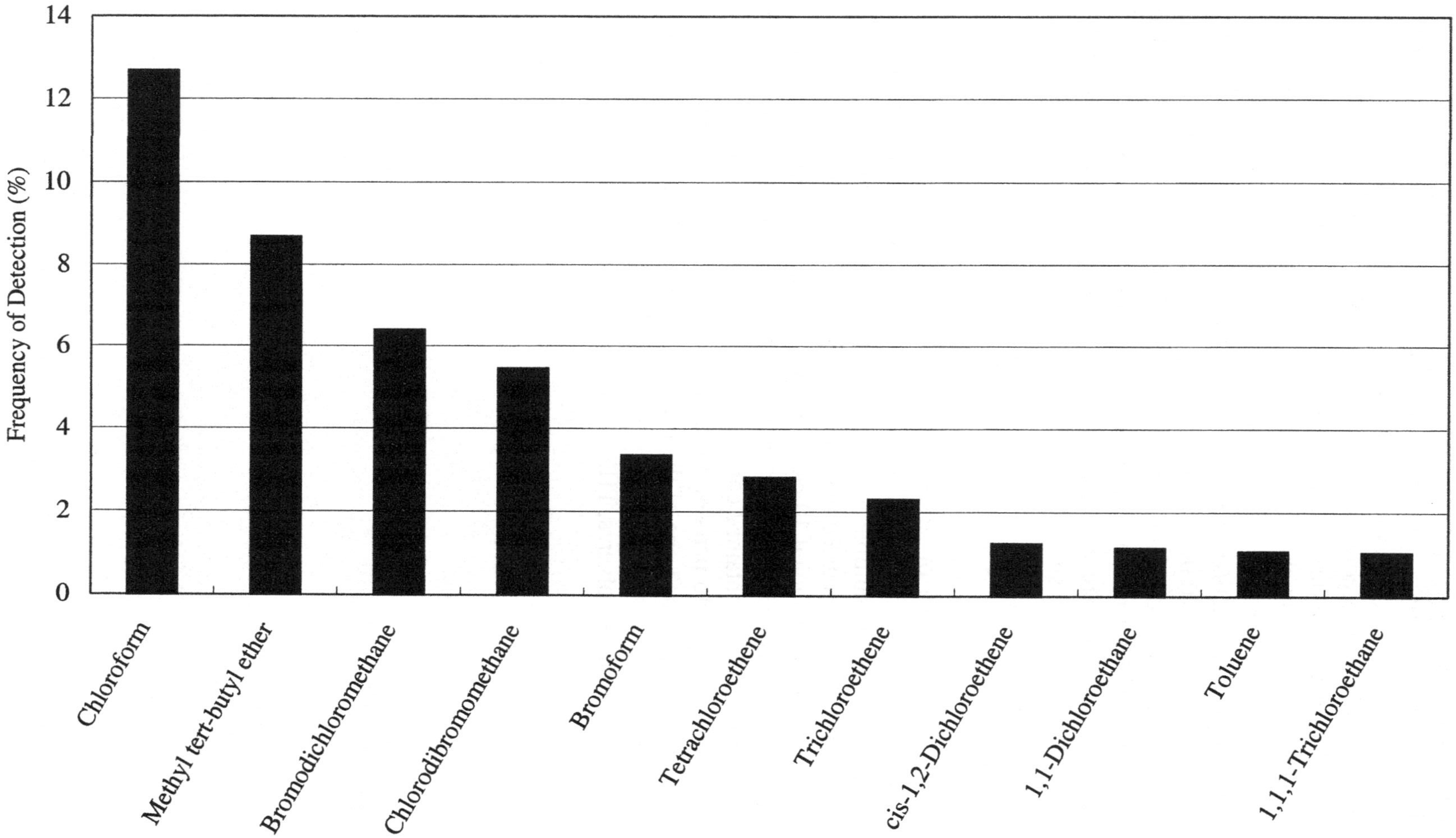

Figure 5.4 Frequency of detection of volatile organic compounds detected in more than 1.0 percent of source-water samples - Random Survey

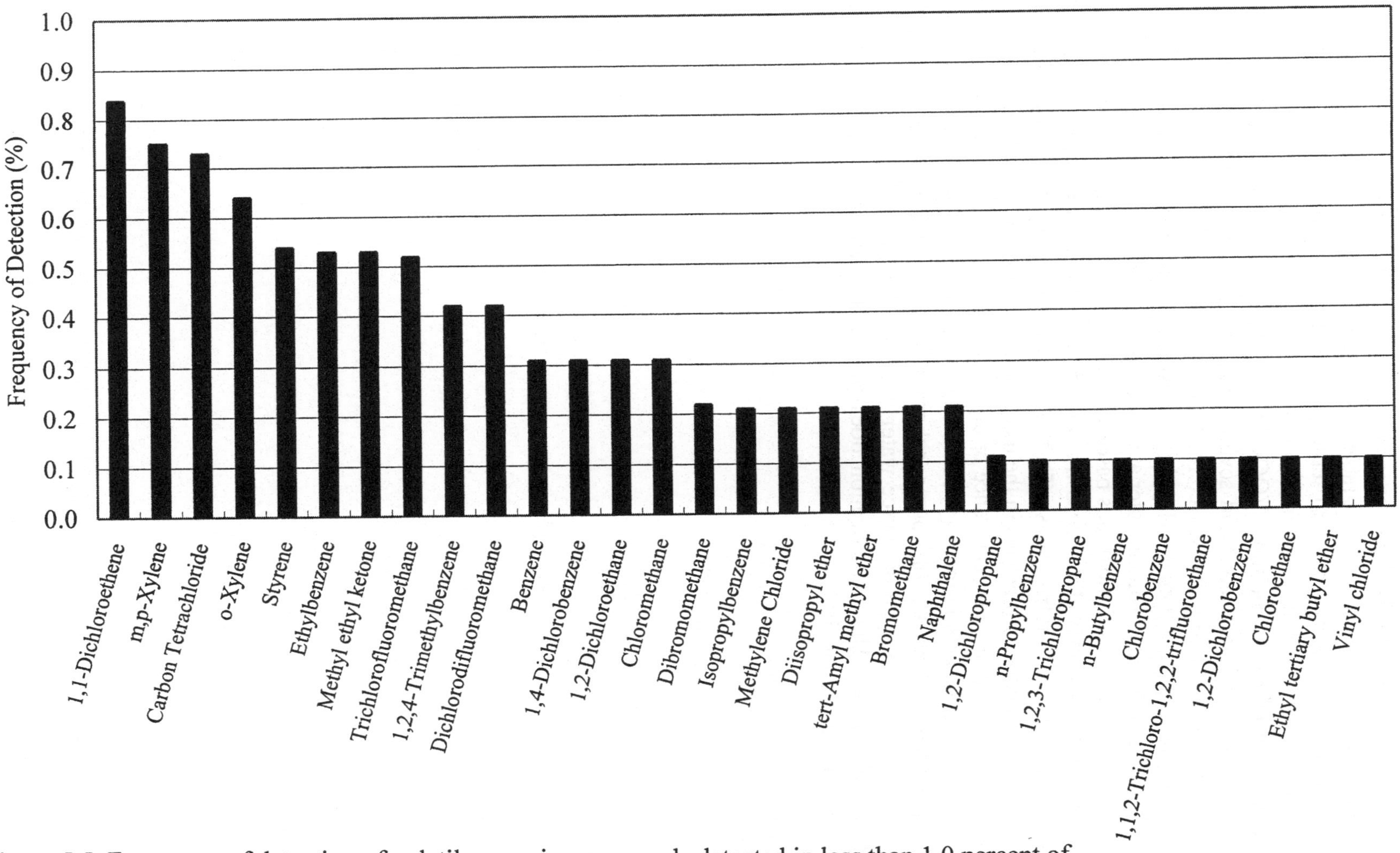

Figure 5.5 Frequency of detection of volatile organic compounds detected in less than 1.0 percent of source-water samples - Random Survey

Co-occurrence can take place when several VOCs have a common source; for example, the presence of several trihalomethanes in a sample may be related to disinfection with chlorine, or the co-occurrence of several BTEX compounds suggests a gasoline source. A high percentage of co-occurrence also can reflect the degradation of parent VOC compounds to their by-products, for example, the co-occurrence of tetrachloroethene and trichloroethene in 83 and 91 percent, respectively, of the source-water samples that contained cis-1,2-dichloroethene. By contrast, the co-occurrence of multiple VOCs from different subgroups, including MTBE, chloroform and other trihalomethanes, and several common solvents in source-water samples does not necessarily imply a common source for these contaminants. Rather, their frequent co-occurrence may be an artifact of their overlapping widespread occurrence.

The 42 VOCs detected in source water at concentrations equal to or greater than the MRL are classified in Table 5.4 into six subgroups based on their most common use or probable source relative to their occurrence in water. These include (1) disinfection by-products, (2) gasoline compounds, (3) solvents, (4) refrigerants, (5) VOCs used in the synthesis of other organic chemicals, and (6) fumigants. Similarities in the uses and (or) sources of VOCs in each of the six subgroups may be reflected in their occurrence or co-occurrence and their distribution. Although most VOCs have numerous uses in industry, commerce, and household applications, the purpose of identifying a predominant use or source subgroup is to facilitate comparisons of their occurence and distribution by type of source water, size of CWS, as well as by anthropogenic factors later in this report. This classification does not imply that other uses or sources for specific compounds do not exist.

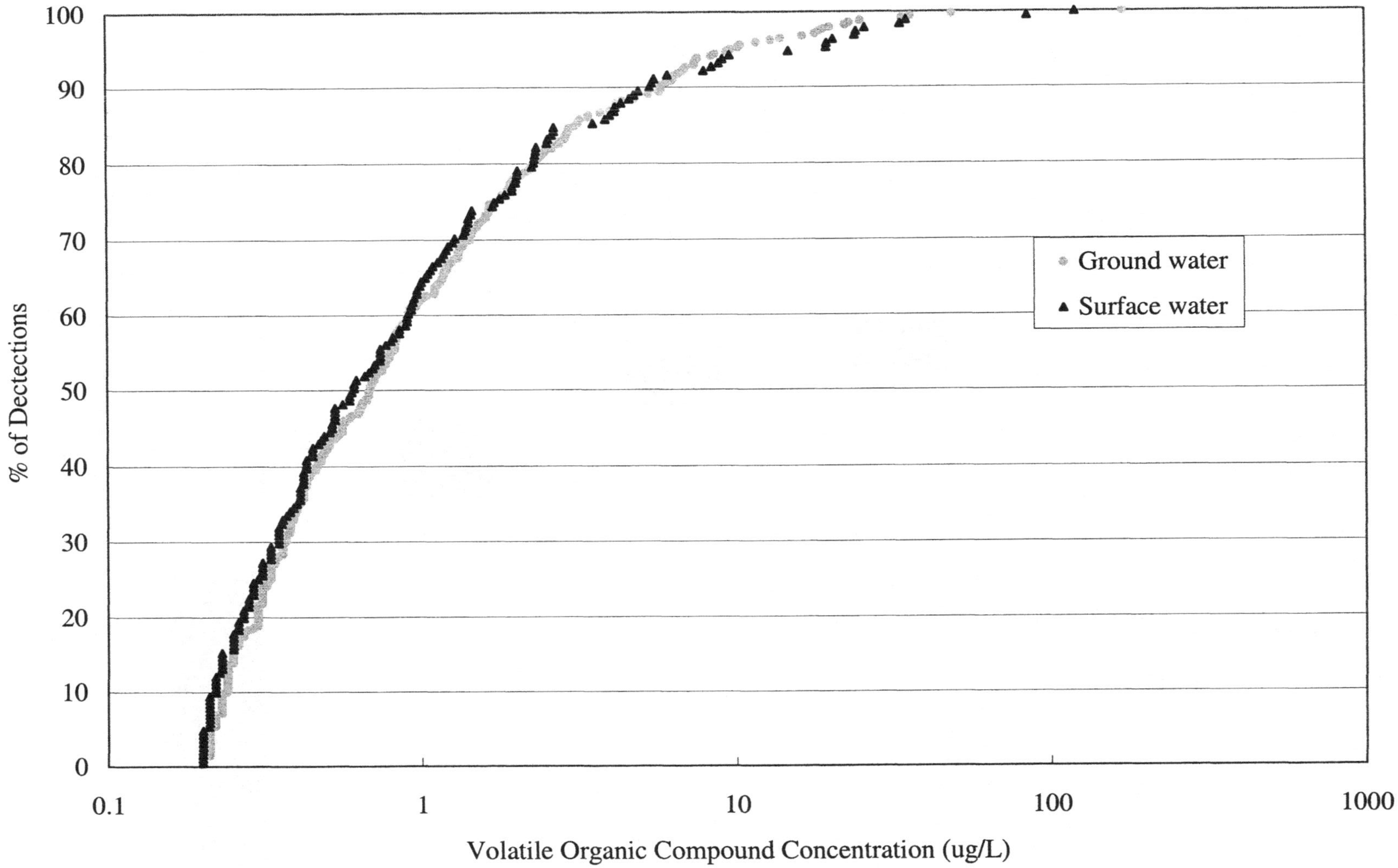

Figure 5.6 Cumulative distribution of concentrations of volatile organic compounds detected in source-water samples from ground-water and surface-water sources - Random Survey

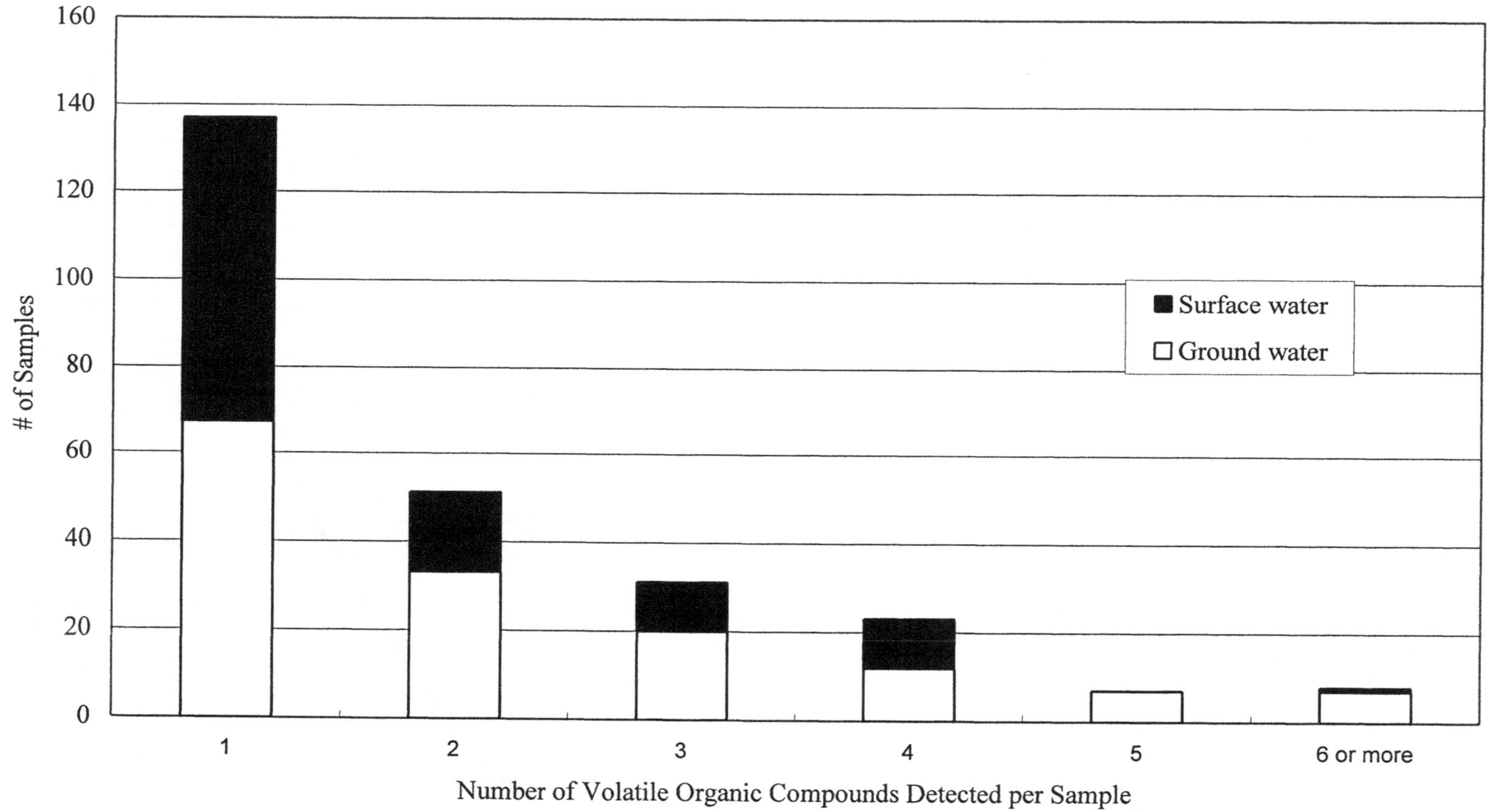

Figure 5.7 Number of volatile organic compounds detected in source-water samples from ground-water and surface-water sources - Random Survey

Table 5.4
Chemical subgroups of volatile organic compounds detected in source-water samples – Random Survey

Chemical use or source subgroup	Compounds detected (listed in order of decreasing detection frequency)	Number of detections	Number (percent) of samples with detections
Disinfection by-products	Chloroform, bromodichloromethane, chlorodibromo- ethane, bromoform	265	141 (15)
Gasoline compounds	Methyl *tert*-butyl ether, toluene, *m*-, *p*-xylene, *o*-xylene, ethylbenzene, benzene, diisopropyl ether, *tert*-amyl methyl ether, naphthalene, ethyl *tertiary* butyl ether, *n*-butylbenzene	120	94 (9.8)
Solvents	Tetrachloroethene, trichloroethene, *cis*-1,2-dichloro- ethene, 1,1-dichloroethane, 1,1,1-trichloroethane, 1,1-dichloroethene, carbon tetrachloride, methyl ethyl ketone, 1,2-dichloroethane, dibromomethane, methyl ene chloride, 1,2-dichloropropane, chlorobenzene, chloroethane, 1,2-dichlorobenzene, *n*-propylbenzene, 1,2,3-trichloropropane	115	60 (6.3)
Refrigerants	Trichlorofluoromethane, dichlorodifluoromethane, chloromethane, 1,2,2-trichloro-1,1,2-trifluoroethane	13	10 (1.0)
Organic synthesis compounds	Styrene, 1,2,4-trimethylbenzene, isopropylbenzene, vinyl chloride	12	10 (1.0)
Fumigants	1,4-dichlorobenzene, bromomethane	5	5 (0.5)

Comparison by Type of Source Water and Size of Community Water System

Proportionally, slightly more surface-water samples (30 percent) than ground-water samples (25 percent) contained VOCs, but the differences in detection frequency by type of source water were not statistically significant for most VOC subgroups (Table 5.5). A contingency-table analysis showed that gasoline compounds in general, and MTBE in particular, were detected significantly more frequently in surface-water samples than ground-water samples. BTEX compounds also were detected more frequently in surface water than in ground water, but the number of BTEX detections were too few to demonstrate a statistically significant relation with source-water type. The elevated detection frequencies for gasoline-related VOCs in surface waters may relate to watercraft use on the affected water bodies. Because most concentrations of gasoline compounds were small, atmospheric and (or) storm-water runoff sources also may have contributed to the more frequent detections (Delzer et al. 1996; Lopes and Bender 1998; Bender et al. 2000). Incidents of contamination of ground water from solvents are well documented, and these compounds were detected more frequently in samples from ground-water sources than from surface-water sources in the Random Survey, as were the rarely detected chlorofluorocarbon refrigerants. Generally, the number and variety of compounds detected were greater for ground-water samples than for surface-water samples. VOCs would likely be subjected to faster and more effective dilution and removal processes in surface water than in ground water.

Table 5.5
Statistical comparison of the occurrence of volatile organic compounds in source-water samples by type of source-water – Random Survey

Volatile organic compound, subgroup, or related compounds	p-value	Type of source water and frequency of detection (population)	
		Percentage of surface-water samples	Percentage of ground-water samples
Any volatile organic compound	0.1567	30 (A)	25 (A)
Trihalomethanes	0.6086	14 (A)	16 (A)
Non-trihalomethane compounds	0.113	18 (A)	14 (A)
Any gasoline compound	<0.0001*	15 (B)	6.6 (A)
Methyl *tert*-butyl ether	<0.0001*	14 (B)	5.4 (A)
BTEX compounds	0.4877	2.4 (A)	1.6 (A)
Solvents	0.0001*	2.4 (A)	8.8 (B)
Refrigerants	0.0256*	0 (A)	1.7 (B)
Fumigants	0.6236	0.8 (A)	0.3 (A)
Organic synthesis compounds	0.3071	1.6 (A)	0.7 (A)

* - significant difference at 95 percent confidence level
p-value - the attained significance level
< - less than
source-water type populations that share the same letter (A, B) are not significantly different
BTEX - benzene, toluene, ethylbenzene, and *m*-, *o*- and *p*-xylenes

For most VOC subgroups, detections were more frequent in samples from the very large water systems (Table 5.6), particularly for ground-water sources. A contingency-table analysis of detection frequencies in surface-water sources by size category, however, found few significant differences at the 95-percent confidence level. Overall, the frequency of detecting any VOC at or above the MRL was greatest (42 percent) in ground-water sources used by the very large systems (Table 5.6) followed by (33 percent) surface-water sources for medium-sized systems. However, VOCs were detected in nearly 20 percent or more of source-water samples for all source-size categories. Although detections of any VOC collectively were significantly more frequent in ground-water sources of very large CWSs than in ground-water supplying very small, small, and medium-sized CWSs, the p-value for this comparison is even more significant when the four trihalomethane compounds are removed and only non-trihalomethane compound detections are compared (Table 5.6). Detections of any gasoline compound, solvent, and refrigerant also were significantly more frequent in ground-water sources to very large CWSs (Table 5.6) than in ground-water sources of smaller-sized CWSs.

The more frequent occurrence of VOCs in source waters used by larger CWSs likely reflects the greater proximity of drinking-water sources for larger systems to high-population density, urban areas compared to the locations of drinking-water sources for the smaller CWSs (Figure 5.8). Urban land use and high-population density have been shown to be related to increased VOC detection frequencies in a number of previous studies (Reiser and O'Brien 1989; Grady 1994; Grady and Mullaney 1998; Squillace et al. 1999; Bush et al. 2000). The relations between the occurrence of VOCs and population density and urban land use are evaluated further in a later section of this chapter.

Trihalomethane disinfection by-products. Four trihalomethane compounds—chloroform, bromodichloromethane, chlorodibromomethane, and bromoform—are among the by-products commonly generated by the disinfection of drinking-water supplies with chlorine and (or) bromine. The four THM compounds were among the five most frequently detected VOCs in the source-water samples (Figure 5.4) and, collectively, were detected in 14 percent of the surface-water sources and in 16 percent of the ground-water sources (Table 5.5). Detections of chloroform and bromodichloromethane (Figure 5.9) were slightly more common in source-water samples from rivers than in samples from ground-water or reservoirs, but detection frequencies of all four THMs were comparable (that is, not significantly different) among the three source- water types. Furthermore, the frequency of detection of THMs was not particularly related to CWS size—proportionally, just as many source-water samples from very small ground-water systems had THM detections as did samples from the very large ground-water systems (Table 5.6), and medium-sized surface-water CWSs reported the most frequent THM detections among the surface-water sources.

Table 5.6
Statistical comparison of the occurrence of volatile organic compounds in source-water samples by type of source water and size of community water systems.

Volatile organic compound or related compounds	Type of source water	p-value	Size of CWSs and frequency of detection (population)				
			Very small	Small	Medium	Large	Very large
Any volatile organic compound	Ground water	0.0381*	26 (A)	20 (A)	20 (A)	24 (AB)	42 (B)
	Surface water	0.7714	25 (A)	19 (A)	33 (A)	29 (A)	31 (A)
Trihalomethanes	Ground water	0.4030	18 (A)	13 (A)	12 (A)	9.5 (A)	17 (A)
	Surface water	0.4635	8.3 (A)	16 (A)	24 (A)	16 (A)	12 (A)
Non-THM volatile organic compounds	Ground water	0.0001*	11 (A)	12 (A)	10 (A)	19 (AB)	36 (B)
	Surface water	0.2190	17 (A)	3.8 (A)	17 (A)	15 (A)	22 (A)
Any gasoline compound	Ground water	0.0466*	5.8 (A)	5.0 (AB)	2.0 (A)	9.5 (AB)	15 (B)
	Surface water	0.1414	8.3 (A)	0 (A)	13 (A)	14 (A)	18 (A)
Methyl *tert*-butyl ether	Ground water	0.0511	4.8 (A)	3.4 (A)	2.0 (A)	8.2 (A)	13 (A)
	Surface water	0.1054	8.3 (A)	0 (A)	6.7 (A)	13 (A)	17 (A)
BTEX compounds	Ground water	0.5642	1.7 (A)	0.8 (A)	0 (A)	1.6 (A)	3.8 (A)
	Surface water	0.0678	0 (A)	0 (A)	10 (A)	2.5 (A)	1.8 (A)
Solvents	Ground water	<0.0001*	4.1 (A)	8.3 (AB)	8.0 (AB)	14 (BC)	30 (C)
	Surface water	0.2687	8.3 (A)	0 (A)	6.7 (A)	2.5 (A)	1.8 (A)
Refrigerants	Ground water	0.0002*	1.4 (A)	0 (A)	2.0 (AB)	0 (AB)	9.4 (B)
	Surface water	1.0000	0 (A)	0 (A)	0 (A)	0 (A)	0 (A)
Fumigants	Ground water	0.3572	0.3 (A)	0 (A)	0 (A)	0 (A)	1.9 (A)
	Surface water	0.7450	0 (A)	0 (A)	0 (A)	0 (A)	1.3 (A)
Organic synthesis compounds	Ground water	0.0619	0.7 (A)	0 (A)	0 (A)	0 (A)	3.8 (A)
	Surface water	0.0245*	8.3 (AB)	3.8 (AB)	6.7 (B)	1.3 (AB)	0.4 (A)

p-value - the attained significance level
* significant difference at 95 percent confidence level
source-size populations that share the same letter (A,B) are not significantly different;
CWSs - community water systems;
BTEX - benzene, toluene, ethylbenzene, and *m*-, *o*- and *p*-xylenes

Because the Random Survey specifically targeted untreated source waters, the widespread occurrence of THMs was not anticipated. The presence of THMs and other disinfectant by-products in finished drinking water has been well documented (Westrick 1990; Minear and Amy 1995; Pomes et al. 1999). Grady and Casey (2001) reported that drinking water supplied by nearly half of all CWSs surveyed in that study contained one or more of the four THMs. Indeed, drinking-water suppliers are required to maintain sufficient chlorine residual levels in their water-distribution systems to prevent the existence of waterborne pathogens. It is possible that samplers from some of the participating CWSs misinterpreted the sample collection instructions and collected finished-water samples rather than source-water samples. Alternatively, some CWSs report that they back flushed their filters with chlorinated water that was discharged back into the source waters. The cause of the anomalous occurrence of concentrations of the total of the four trihalomethane compounds (THMs) in excess of 20 μg/L in nine source-water samples and in excess of the 80-μg/L MCL in one sample (Figure 5.10) is undetermined.

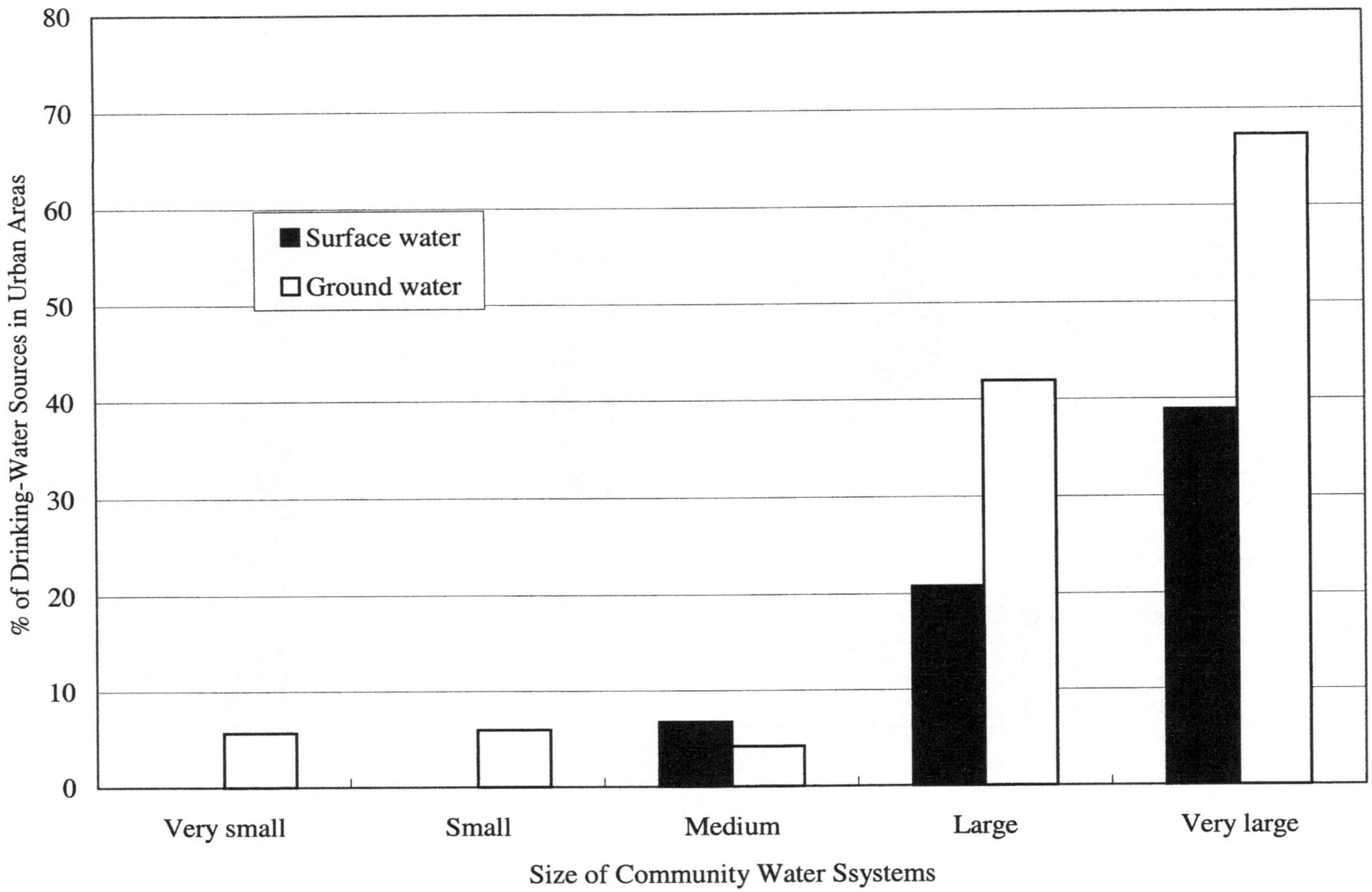

Figure 5.8 Relation between size of community water systems and percentage of drinking-water sources in urban areas - Random Survey

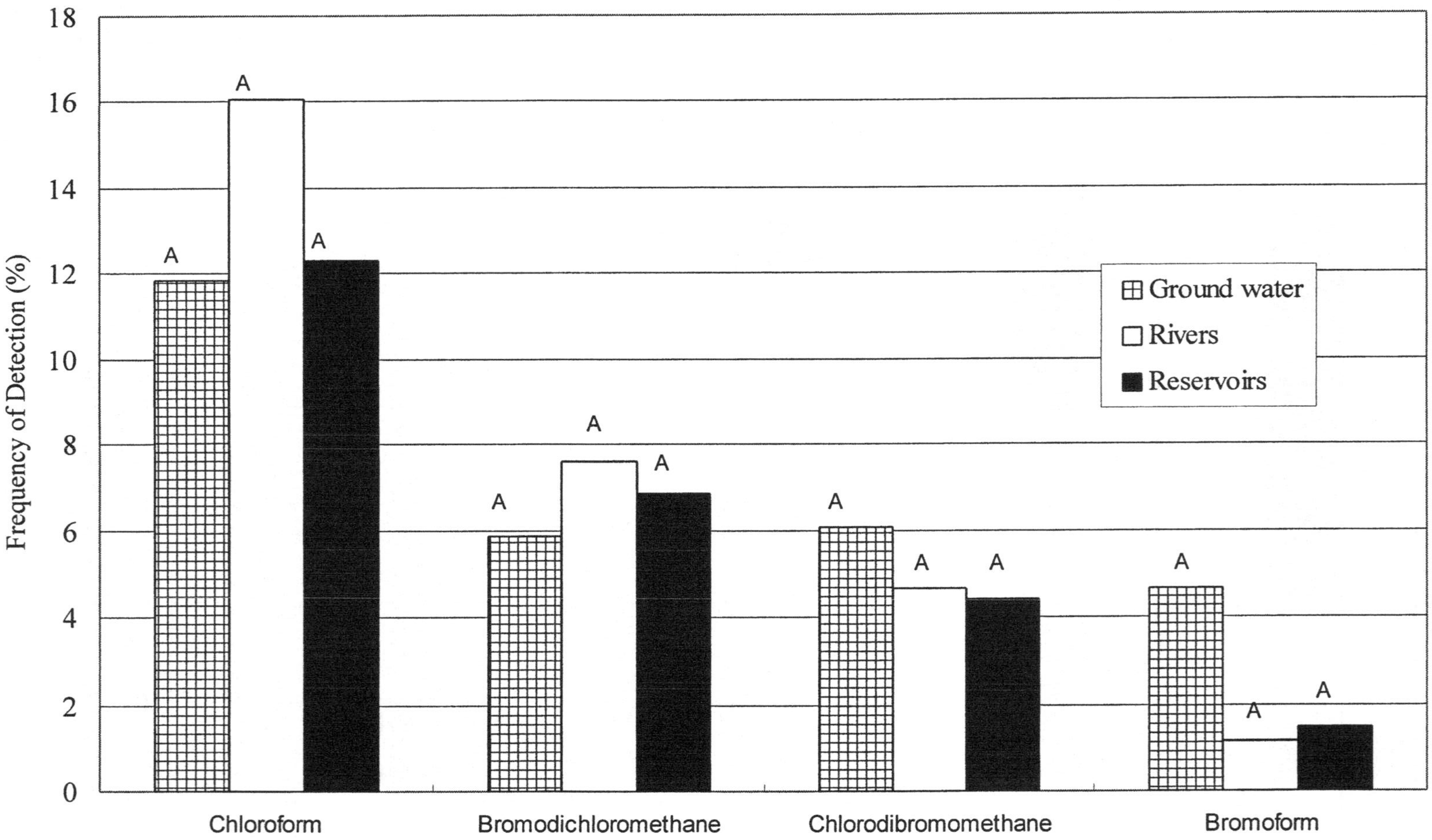

Note - populations that share the same letter (A,B) are not significantl
at the 95-percent confidence level

Figure 5.9 Frequency of detection of trihalomethanes in source-water samples from ground water, rivers and reservoirs - Random Survey

Chloroform in source waters could originate from sources other than a by-product of disinfection, including manufacture of pharmaceuticals, dry cleaning, fire extinguishers, and fumigants. However, chloroform frequently co-occurred in source-water samples with other THMs in the following proportions: chloroform > bromodichloromethane > chlorodibromomethane > bromoform. THMs in chlorinated water typically are present in these same proportions. An explanation for widespread low-level THM detections in source waters is that they may represent residual concentrations of disinfectant by-products that formed initially at water or wastewater treatment plants. Disinfectant by-product residuals persist in the environment from irrigation of lawns and parks, sewer exfiltration, and treatment-plant effluents. They circulate in surface water and ground water until intercepted by water-supply intakes and are detected in the source-water samples. Ambient ground water was reported to frequently contain THMs, particularly chloroform, that was attributed to infiltration of treated water used to water lawns or from leaky water and sewer lines, but median concentrations were less than 1.0 μg/L (Squillace et al. 1999). THM concentrations measured in the source-water samples were similar, and were not significantly different among the three source-water types (Figure 5.10). The median concentration of THMs in ground-water sources (0.86 μg/L) was comparable to the 2.5-μg/L median reported by Grady and Casey (2001) for finished drinking water from ground-water supplied CWSs. Median concentrations of THMs in surface-water sources (Figure 5.10), however, were substantially lower than the elevated median TTHM concentrations of 20 to 30 μg/L reported by Grady and Casey (2001) in finished drinking water from surface-water supplied CWSs. The temporal distribution of THM detections also was examined to determine if any seasonal patterns were evident for any particular source-water type. Although substantial differences in the frequency of THM detections from month to month were found in both ground-water and surface-water samples, these variations appear to be random. There was no systematic seasonal pattern evident in the data (Figure 5.11).

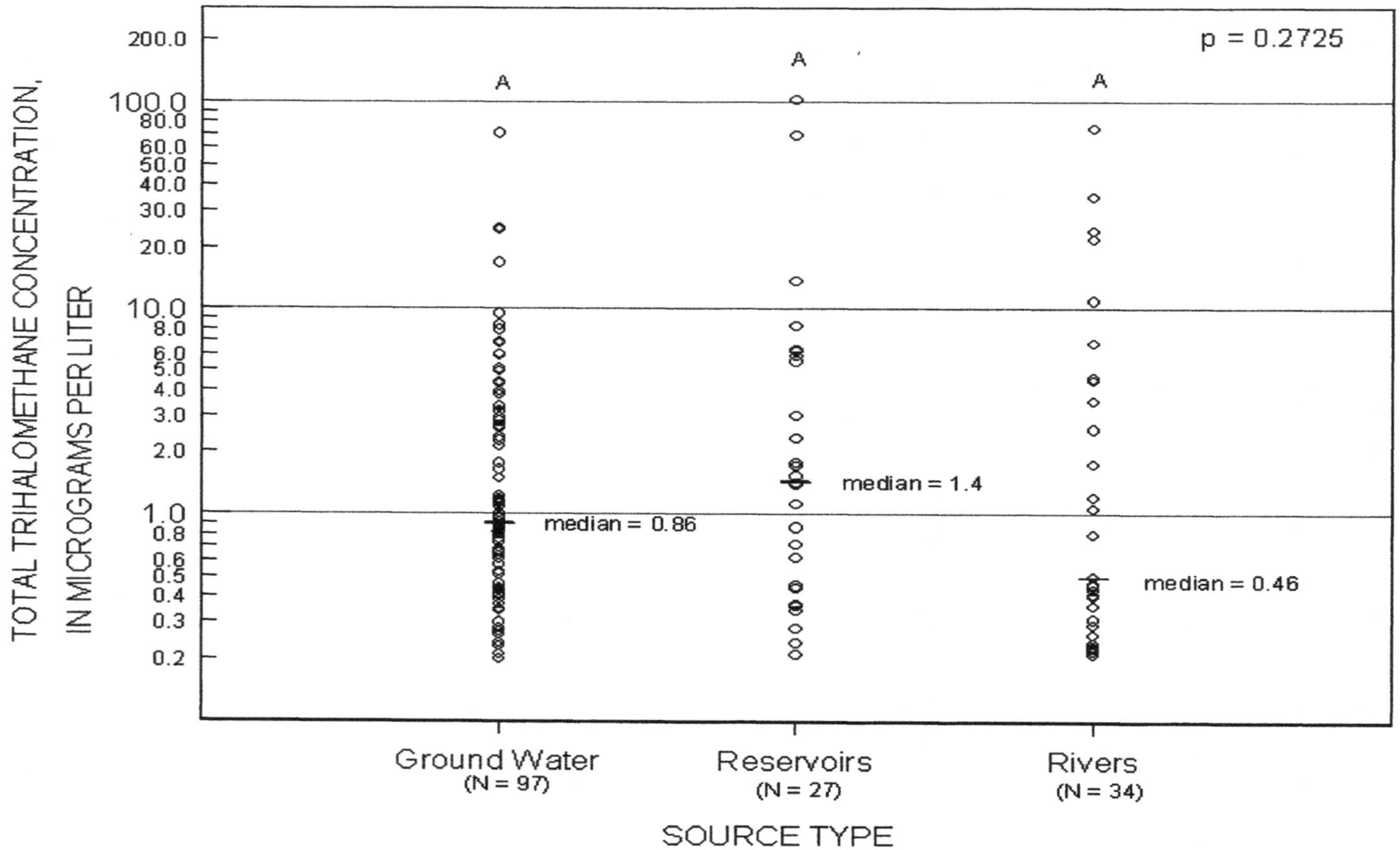

N= number of detections, p = attained significance level; populations that share the same letter symbol are not significantly different at the 95% confidence level

Figure 5.10 Concentrations of total THMs in source-water samples from groundwater, rivers, and reservoirs – Random Survey

MTBE and other gasoline compounds. Gasoline compounds were the second most frequently detected subgroup of VOCs and were measured in 9.8 percent of the source-water samples (Table 5.4). MTBE alone accounted for two-thirds of all gasoline-compound detections (8.7 percent of the source-water samples), and was the second most frequently detected VOC in source waters after chloroform (Figure 5.4). Ten additional VOCs that are intrinsic components of gasoline, such as benzene, toluene, ethylbenzene, and xylenes (BTEX), or are additives to gasoline (including the oxygenates ETBE, DIPE, and TAME), also were detected in source-water samples (Table 5.4).

MTBE detections were significantly more frequent (Table 5.5) in surface-water sources (14 percent) than in ground-water sources (5.4 percent), and river and reservoir samples contained MTBE more than twice as often as ground water (Figure 5.12). The occurrence of MTBE in 5.4 percent of the Random Survey ground-water sources matches that observed in the national assessment of ambient ground water by Squillace et al. (1999), also conducted at the same 0.2 μg/L-MRL. However, Grady and Casey (2001) observed more frequent detections of MTBE in finished drinking water from CWSs in the Northeast. In that study, 7.8 percent of CWSs that used ground-water sources exclusively reported MTBE in drinking water (at a 1.0-μg/L MRL), whereas only 2.8 percent of these systems that use only surface-water sources had reported MTBE. CWSs in the Northeast that used both ground-water and surface-water sources, however, reported MTBE more frequently, in 16-percent of the multiple-source systems (Grady and Casey 2001). Toluene was the only other gasoline compound detected significantly more frequently in surface water than in ground water ($p=0.0309$), and it was detected six times more frequently in rivers than in ground water (Figure 5.12). Consistent with findings from Grady and Casey (2001), MTBE and other gasoline compounds generally were detected more frequently in the large CWSs regardless of type of source water (Table 5.6).

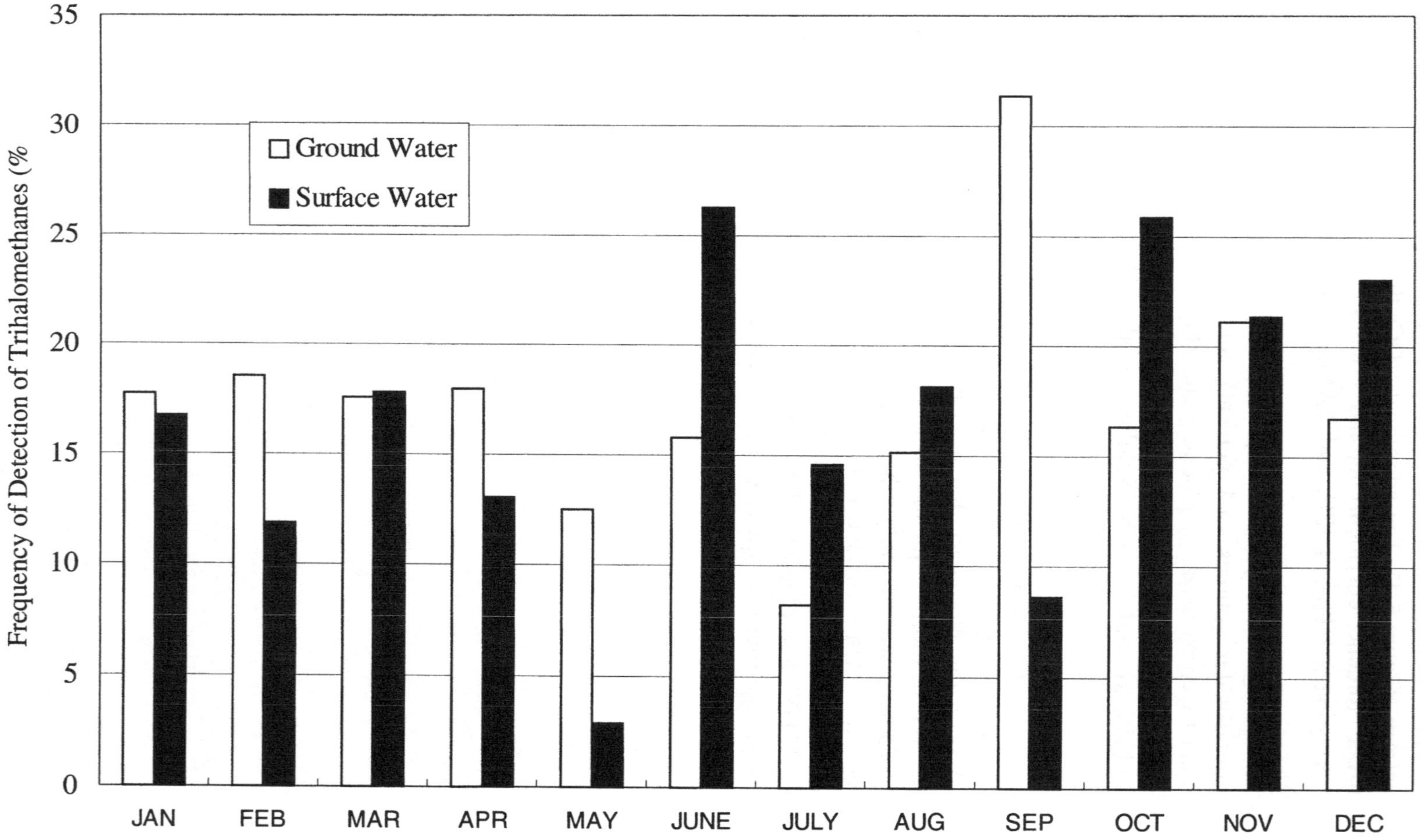

Figure 5.11 Freqency of detection of trihalomethanes in ground- and surface-water samples by month - Random Survey

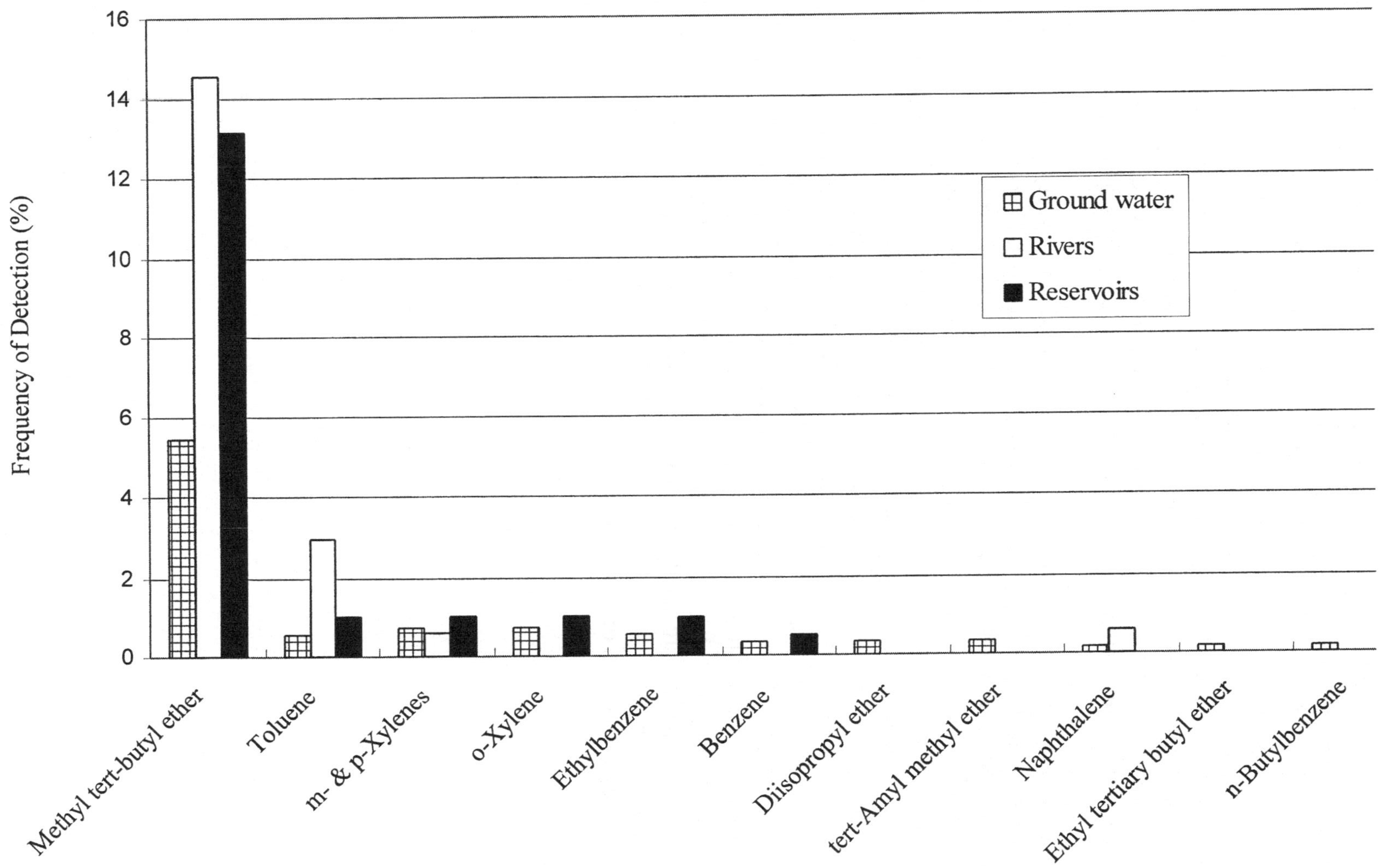

Figure 5.12 Frequency of detection of gasoline compounds in source-water samples from ground water, rivers, and reservoirs - Random Survey

Although widely detected, most MTBE concentrations measured in source-water samples were less than 5.0 μg/L, and median concentrations were less than 1.0 μg/L for each source-water type (Figure 5.13). The highest MTBE concentration measured was 20 μg/L, from a source-water sample (reservoir) in California, and equaled the lower value of the USEPA's recommended drinking-water advisory range for drinking water of 20 to 40 μg/L (U.S. Environmental Protection Agency 1997). The operators of that CWS reported that numerous two-stroke, motorized personal watercraft use the reservoir for recreational purposes. MTBE concentrations did not differ significantly between ground-water and reservoir samples (Figure 5.13), but both were significantly greater than river samples. It is possible that the lower concentrations observed in river source-water samples is because mixing, dilution, and volatilization of gasoline contaminants is more effective in flowing water than stratified water bodies, or it may be that the rivers in this survey collectively receive less motorized watercraft use.

BTEX and other gasoline-related VOCs were detected only rarely in source waters (Figure 5.12) and even more rarely together with MTBE (Figure 5.14). Co-occurring detections of any BTEX compound with MTBE occurred in only 10 of 934 (1.1 percent) source-water samples. When MTBE was detected, BTEX compounds also were detected 12 percent of the time, but nearly 30 percent of the 34 source-water samples that contained BTEX compounds also contained MTBE. Most of the time when these two gasoline components co-occurred, and particularly when the concentrations of MTBE and BTEX were similar (Figure 5.14), the concentrations were low (less than 2.0 μg/L). The few MTBE detections that exceeded 2.0 μg/L were largely without BTEX in source-water samples. The lack of substantial co-occurrence of MTBE and BTEX likely reflects the different chemical properties that cause MTBE to be relatively persistent in the environment compared to BTEX, especially in ground water. Conversely, the ether oxygenates ETBE, DIPE, and TAME, while rare (they were detected only in two ground-water samples), were detected only where MTBE also was detected. The chemical properties of these compounds are similar to those of MTBE, and they are likely present in oxygenated gasoline as impurities or alteration products.

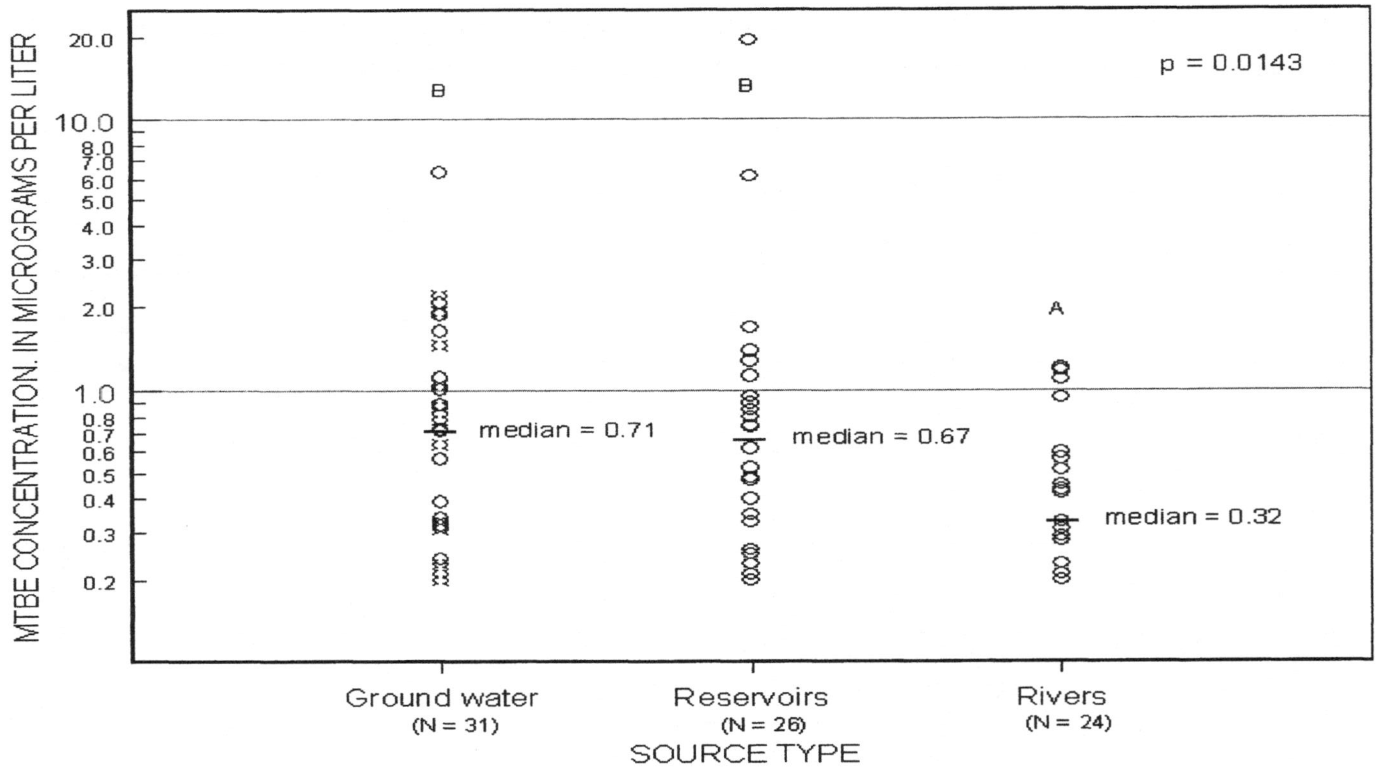

N= number of detections: p = attained significance level; populations that share the same letter symbol are not significantly different at the 95% confidence level

Figure 5.13 Concentrations of MTBE in source-water samples from ground water, rivers, and reservoirs -Random Survey

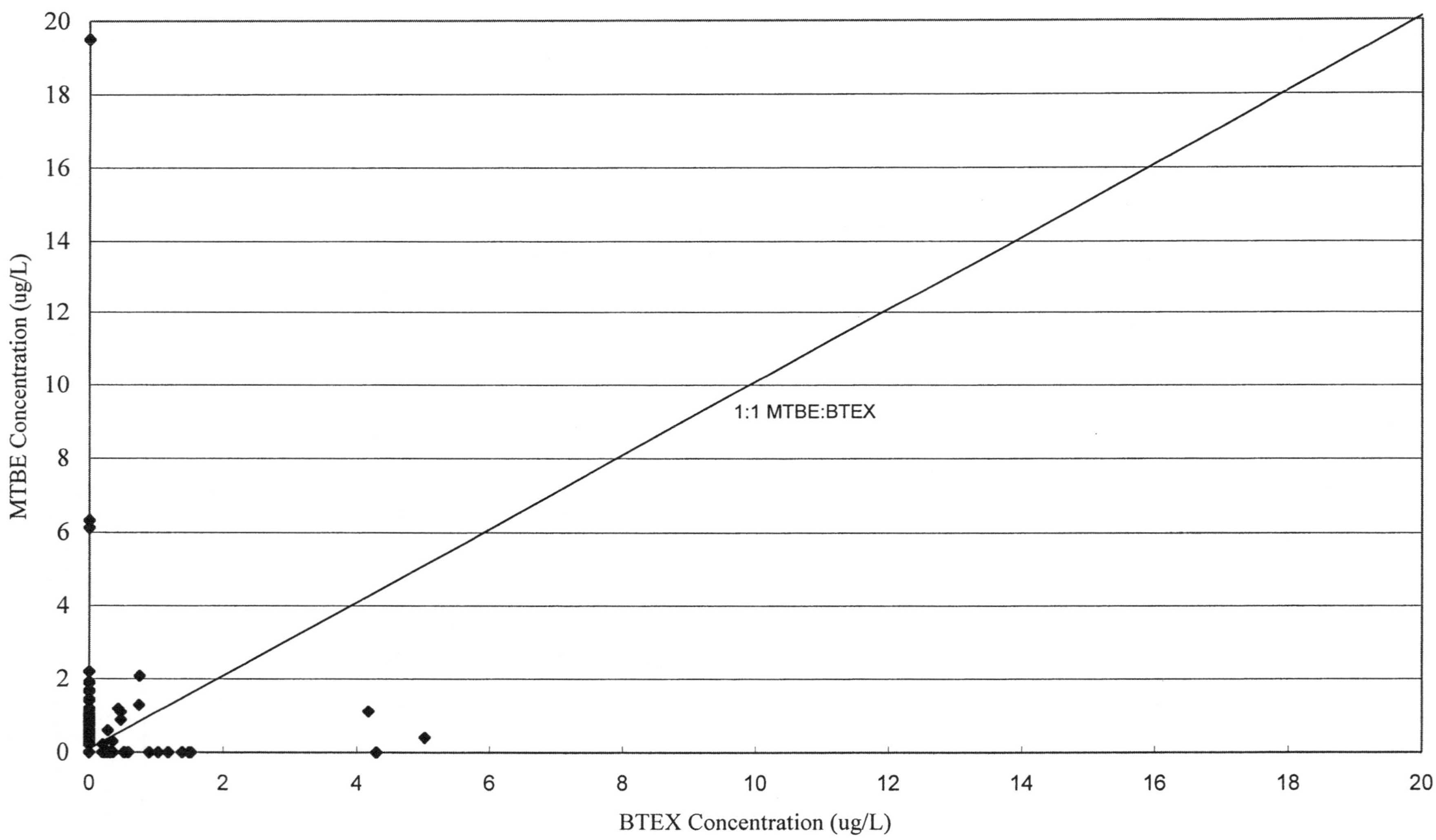

Figure 5.14 Comparison of concentrations of MTBE to the sum of the concentrations of benzene, toluene, ethylbenzene, and xylenes (BTEX) in 934 source-water samples

The temporal distribution of MTBE detections in ground-water sources and surface-water sources (Figure 5.15) demonstrates substantial differences. Although detection frequency varied randomly and no seasonal pattern is evident in the detection of MTBE in ground water, a seasonal influence seems apparent in the surface-water detections. MTBE detection frequency in surface-water samples increases almost monotonically from January to June and then similarly declines through October (Figure 5.15). This pattern may reflect the seasonal use of watercraft on surface-water bodies that are drinking-water sources. Emissions from gasoline-powered watercraft, and from personal recreation watercraft equipped with two-stroke engines in particular, have been associated with MTBE and BTEX contamination of surface waters (Juttner et al. 1995; Boughton and Lico 1998; Reuter et al. 1998; Zapecza and Baehr 1999; Dale et al. 2000; Gabele and Pyle 2000).

Solvents and other volatile organic compounds. Solvents were the third most frequently detected subgroup of VOCs in drinking-water sources, present in 6.3 percent of the samples (Table 5.4). Among the 17 chlorinated solvents detected in source-water samples (Figure 5.16) are several chemicals that have widespread industrial and commercial applications (for example, dry cleaning, textile production, and metal degreasing in automobile-repair and electronic-manufacturing facilities). These compounds also are commonly used in residential areas in household products such as paint strippers, degreasers, aerosols, and adhesives; consequently, solvents have been frequent contaminants in ambient ground water (Squillace et al. 1999) and in drinking water (Grady and Casey 2001).

The occurrence of solvents in drinking-water sources is predominantly a ground-water phenomenon, as 88 percent of all solvent detections were in ground-water samples. Solvents were detected nearly four times more frequently in ground-water sources than in surface-water sources. Detection of one or more solvents in 8.8 percent of the ground-water samples was significantly greater than the 2.4-percent detection frequency for surface-water samples (Table 5.5). Detections of 14 solvents were greater in ground water than rivers and reservoirs (Figure 5.16), and more than half of the 17 solvents detected in source water were present only in ground-water samples. Also, solvents were detected significantly more often in ground-water sources of the largest CWSs (Table 5.6) than in ground-water sources of smaller systems.

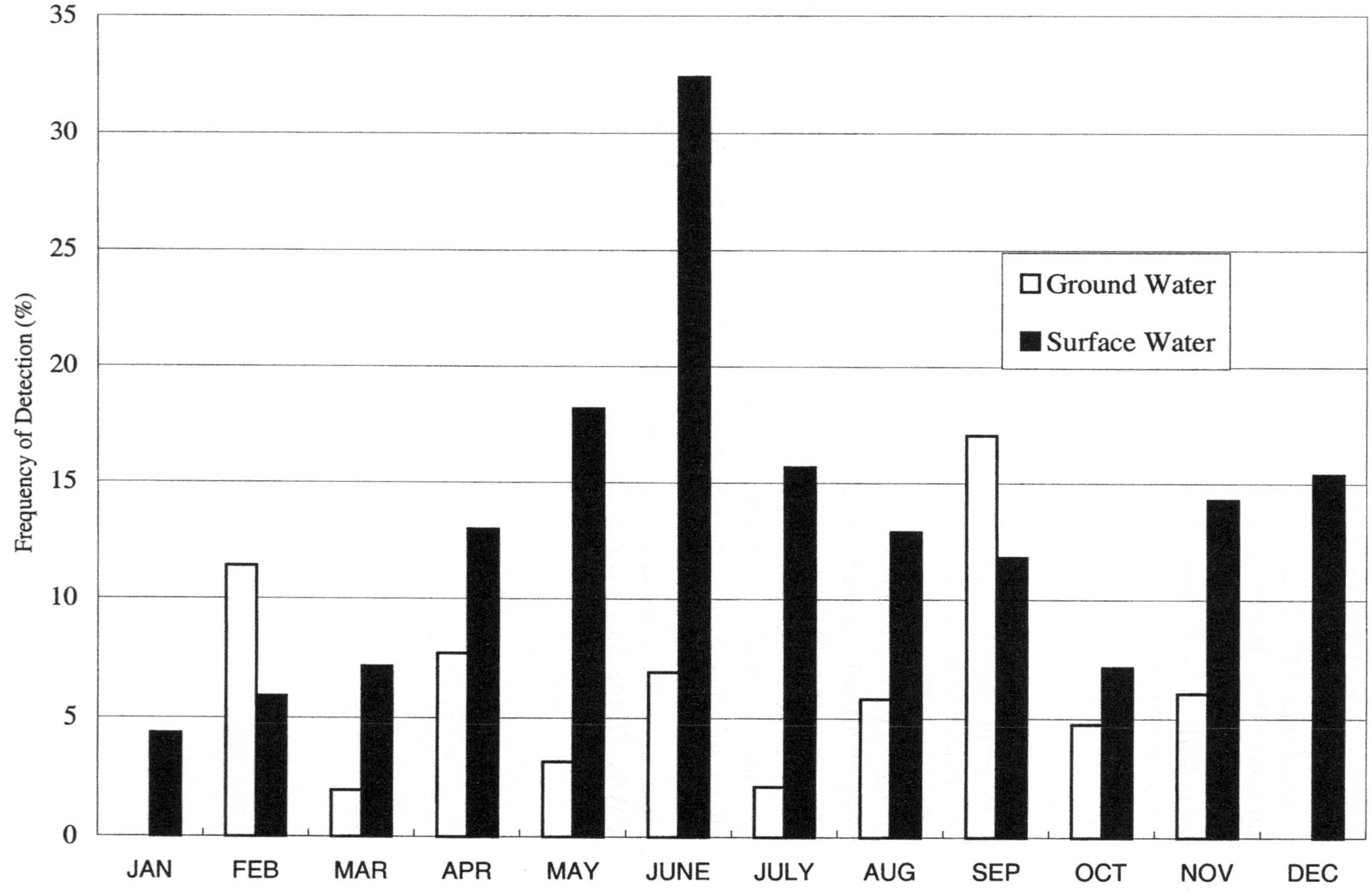

Figure 5.15 Frequency of detection of MTBE in source-water samples from ground-water and surface-water sources by month - Random Survey

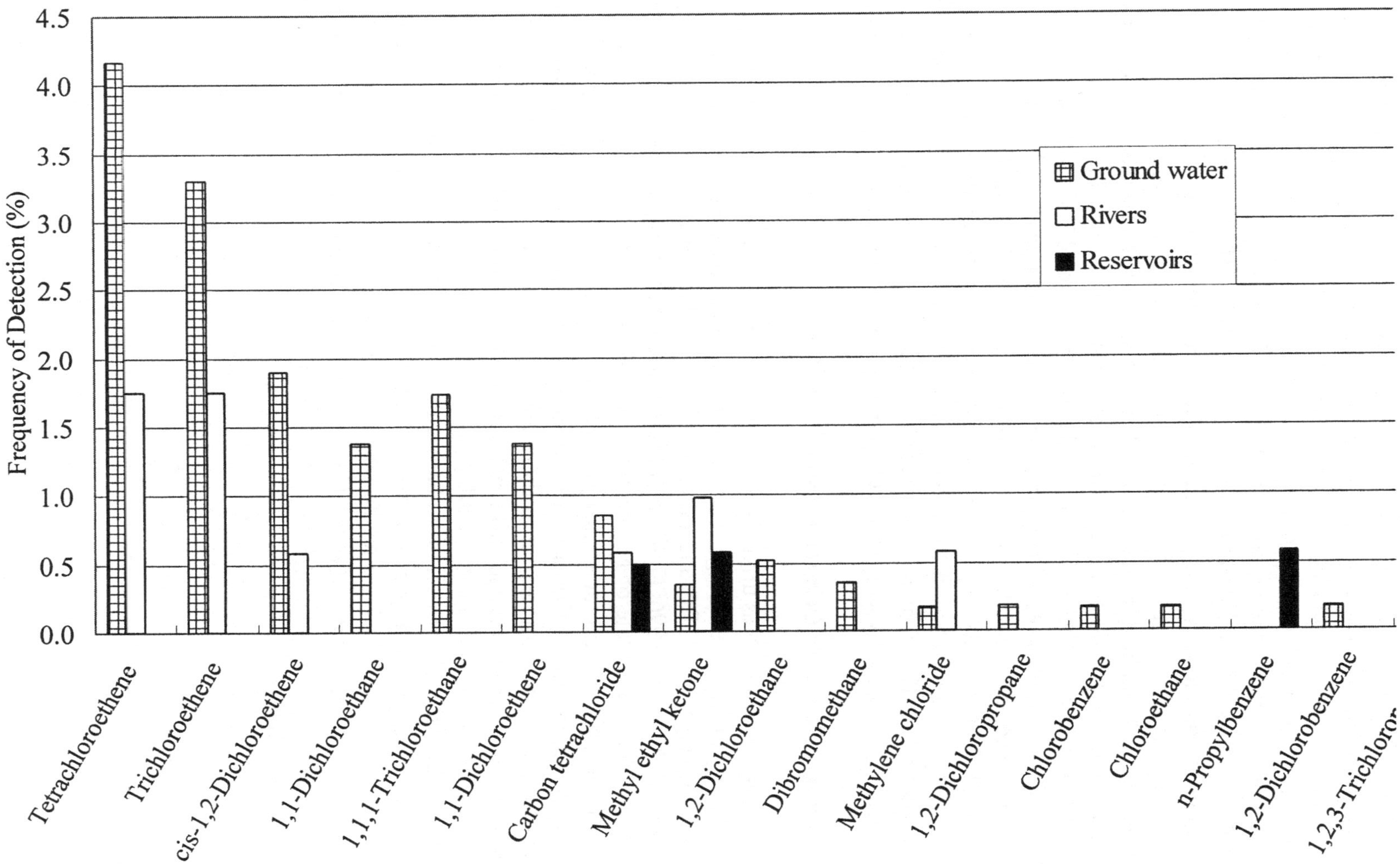

Figure 5.16 Frequency of detection of solvents in source-water samples from ground water, rivers, and reservoirs - Random Survey

Concentrations of solvents in source water were somewhat greater than VOCs in general, as 12 percent of solvent detections exceeded 10 µg/L compared to 5 percent for VOCs overall. Concentrations of tetrachloroethene, trichloroethene, and 1,1-dichloroethene exceeded drinking-water MCLs (5, 5, and 7 µg/L, respectively) 14 times collectively, with a maximum of 165 µg/L of tetrachloroethene measured in one ground-water sample. The samples were not finished drinking water, however, and presumably concentrations of that magnitude would be mitigated by some form of treatment or dilution by mixing with other source waters prior to distribution. At least 15 percent of the CWSs reported using some form of treatment to filter or remove VOCs from source waters sampled for this study (Appendix H – Table H.1). Overall, most solvent concentrations were not problematic, and the median solvent concentration was 1.2 µg/L. Also, perhaps because they were present predominantly in ground water or perhaps because their sources are widespread and random, detections of solvents demonstrated no seasonal patterns.

VOCs other than the trihalomethane disinfectant by-products, gasoline compounds, and solvents described above have been detected only rarely in drinking-water sources. Four refrigerants, four VOCs used mostly in the synthesis of other organic chemicals or products, and two fumigants were detected in less than 1 percent of the source-water samples (Table 5.4). Of the three groups of VOCs, only refrigerants demonstrated a distinctive and statistically significant (see Table 5.5) difference in their occurrence, as they were detected exclusively in ground water (Figure 5.17). The fumigant and organic synthesis VOCs detected in source-water samples were present in all source-water types but were slightly more prevalent in surface water.

Anthropogenic Factors Related to Occurrence and Distribution

Anthropogenic factors such as urban land use, population density, areas where MTBE use is high, and the density of gasoline storage tanks have been found to explain, at least in part, the occurrence and distribution of MTBE and other VOCs in drinking water and ambient ground water (Grady 1994; Delzer et al. 1996; Grady and Mullaney 1998; Hitzig, Kostecki, and Leonard 1998; Lopes and Bender 1998; State of Maine 1998; Moran et al. 1999a; Squillace et al. 1999; Squillace and Moran 2000; Grady and Casey 2001). Consequently, a similar analysis was performed for the Random Survey that used information on the location of the drinking-water sources sampled for this study with respect to these anthropogenic factors. The locations of the drinking-water sources were intersected with national GIS data on 1990 population density (Price and Clawges 1999) and percentage of urban land use (Vogelmann et al. 2001) within the four 1-km2 grid cells adjacent to the well or intake location. Population density was calculated as an average of the four neighboring grid cells weighted for distance of the well or intake from each cell. Urban land use was calculated as the percentage of total area in the four neighboring cells consisting of four land cover types: (1) low-intensity residential, (2) high-intensity residential, (3) commercial, industrial, and transportation, and (4) urban recreational grasses (for example, parks and golf courses).

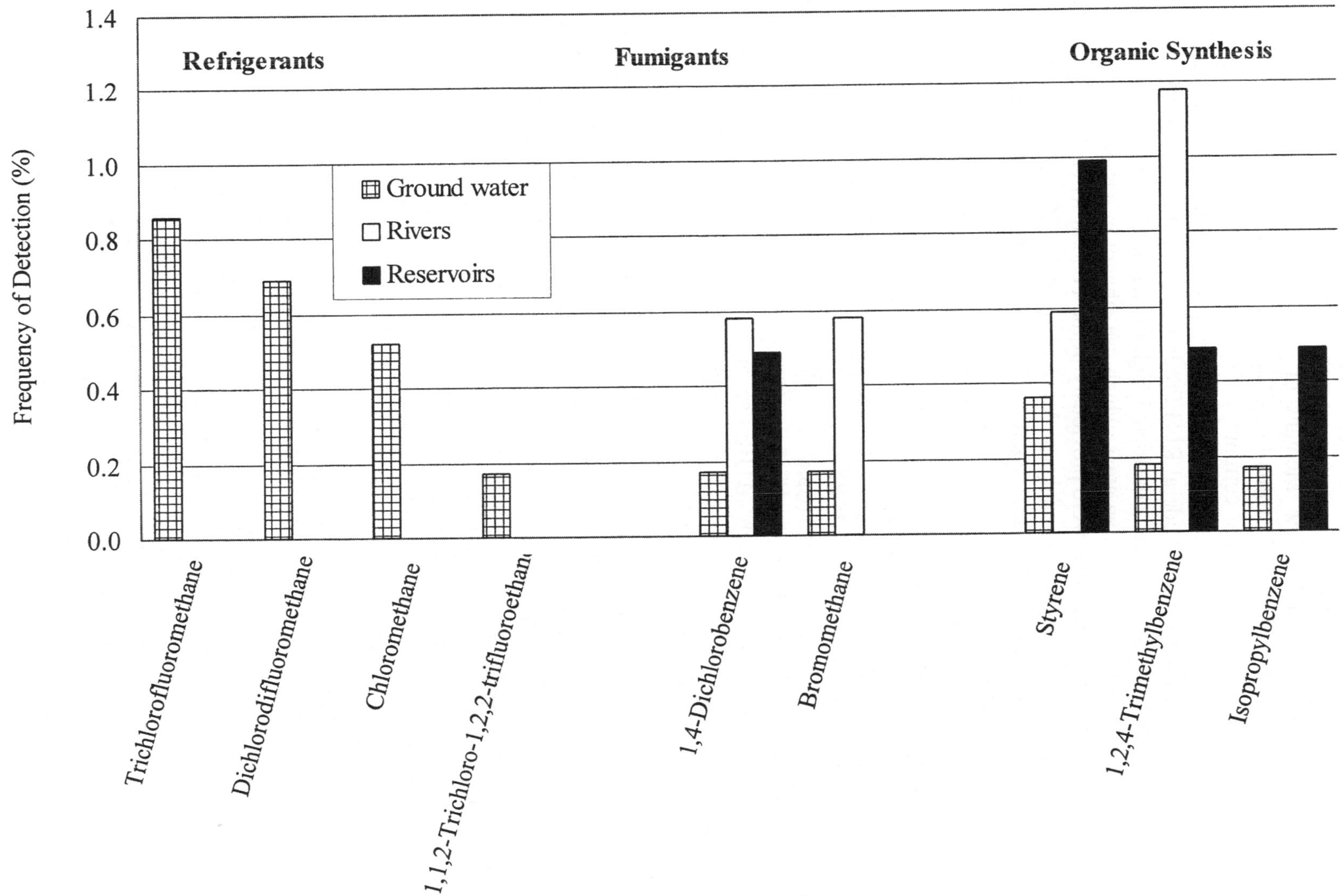

Figure 5.17 Frequency of detection of refrigerants, fumigants, and organic synthesis compounds in source-water samples from ground water, rivers, and reservoirs - Random Survey

High MTBE-use areas were defined as areas within the Federal Reformulated Gasoline (RFG) Program (U.S. Environmental Protection Agency 1999a) whereby gasoline must contain 2-percent oxygen by weight and MTBE is the oxygenate of choice. Areas designated as "high use" had a median content of MTBE in gasoline greater than 9 percent by volume in at least one year or season from 1995 through 1999. Medians of MTBE by volume in gasoline were determined from yearly data from the Reformulated Gasoline Survey (Stuart Romanow, U.S. Environmental Protection Agency, written commun., 1999) and seasonal (winter and summer) data from motor gasoline surveys conducted by the National Institute for Petroleum and Energy Research. All areas participating in the Federal RFG Program during the April 1999 to October 2000 duration of the Random Survey were considered high MTBE-use areas, with the exception of the Chicago and Milwaukee metropolitan areas, which use ethanol in gasoline to meet federal oxygenate requirements. Two areas that were previously in the Federal RFG Program also were considered high MTBE-use areas for this analysis. Areas of southern Maine voluntarily entered (opted into) the RFG Program in 1995 and opted out in 1999. Phoenix, Arizona voluntarily entered the RFG Program in 1997 and opted out in June 1998. Both of these areas have shown significant (>9 percent by volume) MTBE use during 1995-99.

Two other areas also were considered high MTBE-use areas for this analysis, although they did not actively participate in the Federal RFG Program. All of the state of California, including those areas outside RFG Program areas, are considered to be high MTBE-use areas because of documented statewide use of MTBE to meet the state wintertime oxygenate and (or) California Cleaner Gasoline requirements (California Environmental Protection Agency 1997; California Senate 1998; Gomez, Brasil, and Chan 1998). Additionally, Yuma and Mohave counties in Arizona are considered high MTBE-use areas because they reportedly receive gasoline containing high MTBE content from California gasoline distributors (Arizona Department of Environmental Quality 1999).

The three ancillary factors evaluated—population density, urban land use, and high MTBE-use areas—are autocorrelated to some degree. Clearly, areas with high population density (greater than 386 people/km2 or 1,000 people/mi2) are urban, and areas with more than 60 percent total urban land cover generally have high population density. Also, because the Federal RFG Program was directed at improving air quality in places within the country that exceeded air-quality standards, due in large part to emissions from congested automobile use, these areas typically include large urban centers and their surroundings. These ancillary factors are only indirect surrogates for quantitative information on VOC sources, use, and releases to the environment within urban areas that generally is not available. The occurrence and distribution of VOCs in ground water and surface water, however, are related to, and to some degree can be anticipated by, these factors.

Urban land use within the 4-km2 area surrounding 931 of the 954 drinking-water sources (data were not available for Alaska, Hawaii, and Puerto Rico) ranged from 0 to 99.9 percent. For this analysis, sites were segregated into five categories of urban land use corresponding to 0 to 20 percent, 20 to 40 percent, 40 to 60 percent, 60 to 80 percent, and 80 to 100 percent urban. The frequency of detection for various subgroups of VOCs were statistically compared using contingency-table analysis. The results of that analysis demonstrates that the frequency of detection of any VOC, non-THM VOCs, BTEX compounds, solvents, and refrigerants was significantly related to urban land use at the 95-percent confidence level (Table 5.7). For most of the VOC subgroups, the frequency of detection was significantly greater in the 80- to 100-percent urban land-use categories than in the 0- to 20-percent urban areas. The relation between

VOC detections and urban land use can be seen in the comparison of the frequency of detection of non-THM VOCs for the five urban land-use percentage categories (Figure 5.18)—the frequency of detection of non-THM VOCs increased monotonically with increasing urban land use and was significantly greater for all areas with more than 20 percent urbanization.

A complementary analysis of the relation between urban areas and VOC detections in source waters was conducted by comparing the frequency of detection of each VOC subgroup for areas with population density less than and greater than the 1,000-people/mi2 threshold adopted by Squillace et al. (1999) to differentiate urban from rural settings. Squillace et al. (1999) found population density to be a stronger explanatory variable than land use in statistical models that tested various explanatory factors for the frequency of VOC detections in ambient ground water. Similarly to urban land use, the relation between detection frequency of any VOC, non-THM VOC, solvent, and refrigerant and population density is statistically significant (Table 5.8). Unlike urban land use, however, detections of any gasoline compound and MTBE were significantly greater in drinking-water sources from areas of high population density, although BTEX compounds were not. MTBE detections were almost three times more frequent in high population-density areas than in lower population-density areas.

Table 5.7
Statistical comparison of the occurrence of volatile organic compounds in source-water samples by percentage of urban land use – Random Survey

Volatile organic compound or related compounds	p-value	Percentage urban land use* and frequency of detection (population)				
		0 – 20 percent urban	20 - 40 percent urban	40 – 60 percent urban	60 - 80 percent urban	80 - 100 percent urban
Any VOC	0.0323†	24 (A)	29 (AB)	27 (AB)	33 (AB)	42 (B)
Trihalomethanes	0.8277	15 (A)	16 (A)	10 (A)	14 (A)	14 (A)
Non-THM compounds	<0.0001†	11 (A)	17 (B)	19 (BC)	28 (BC)	34 (C)
Any gasoline compound	0.0749	8.3 (A)	11 (A)	12 (A)	19 (A)	14 (A)
Methyl *tert*-butyl ether	0.5055	7.9 (A)	9.0 (A)	9.9 (A)	12 (A)	14 (A)
BTEX compounds	0.0013†	1.1(A)	2.5 (AB)	1.0 (AB)	8.8 (B)	2.0 (AB)
Solvents	<0.0001†	3.4 (A)	7.4 (B)	5.2 (AB)	16 (BC)	26 (C)
Refrigerants	<0.0001†	0.5 (A)	0.6 (A)	1.0 (AB)	0 (AB)	8.0 (B)
Fumigants	0.0912	0.2 (A)	0.6 (A)	2.1 (A)	0 (A)	2.0 (A)
Organic synthesis compounds	0.2317	0.7 (A)	0.6 (A)	0 (A)	3.5 (A)	2.0 (A)

p-value - the attained significance level
urban land-use populations that share the same letter (A,B) are not significantly different
* analyzed for four 1-km^2 grid cells adjacent to the drinking-water source
† significant at 95 percent confidence level
< - less than
BTEX - benzene, toluene, ethylbenzene, and *m*-, *o*- and *p*-xylenes

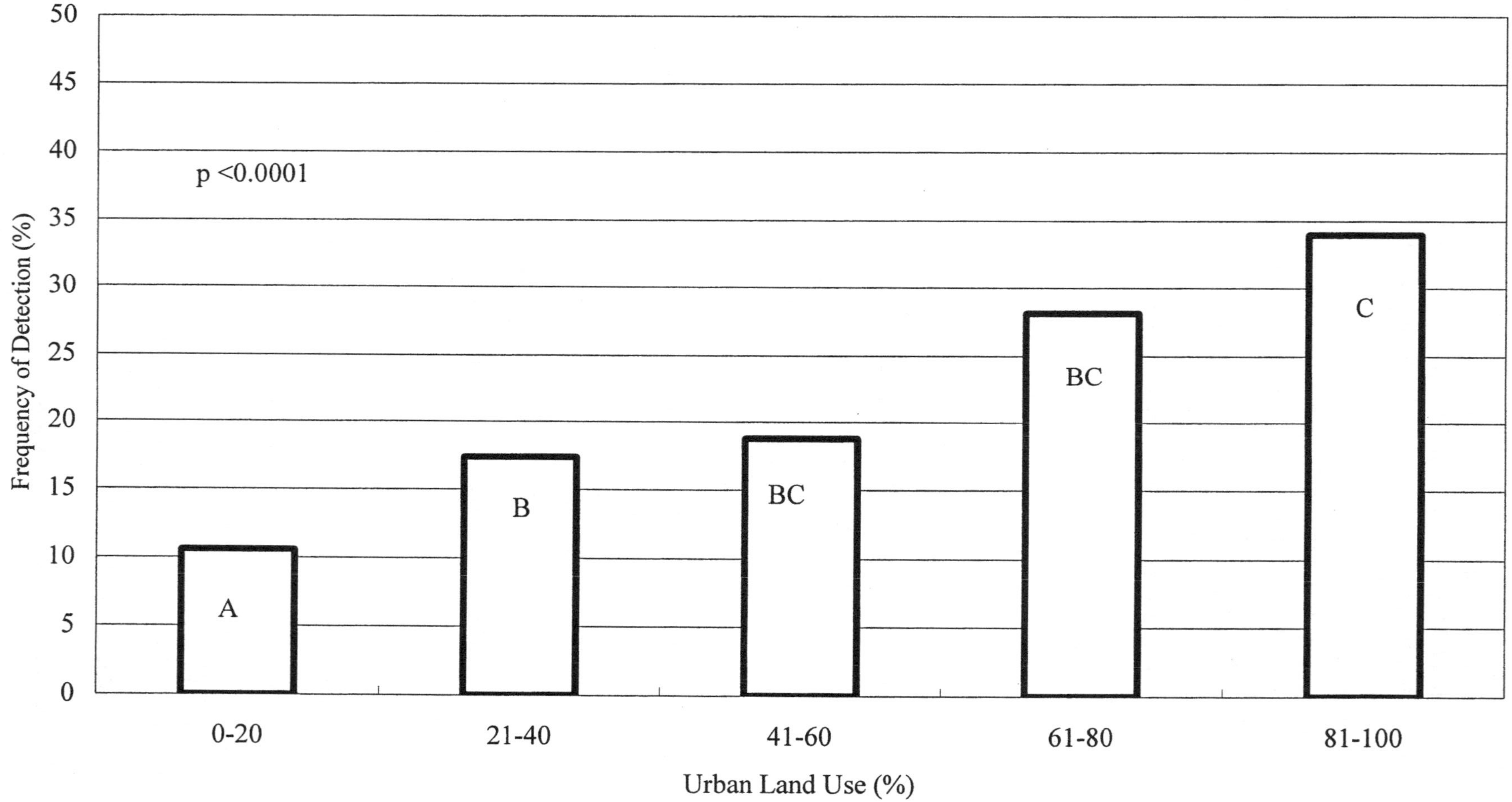

p = attained significance level; populations that share the same letter (A,B,C) are not significantly different

Figure 5.18 Relation between the frequency of detection of non-trihalomethane volatile organic compounds and urban land use - Random Survey

Table 5.8
Statistical comparison of the occurrence of volatile organic compounds in source-water samples by population density.

Volatile organic compound or related compounds	p-value	Population density* and frequency of detection (population)	
		Less than 1,000 people per square mile	Greater than 1,000 people per square mile
Any VOC	0.0003†	24 (A)	38 (B)
Trihalomethanes	0.4921	14 (A)	16 (A)
Non-THM compounds	<0.0001†	13 (A)	28 (B)
Any gasoline compound	0.0001†	8.2 (A)	19 (B)
Methyl *tert*-butyl ether	<0.0001†	7.0 (A)	17 (B)
BTEX compounds	0.2192	1.6 (A)	3.3 (A)
Solvents	0.0009†	4.9 (A)	12 (B)
Refrigerants	0.0020†	0.4 (A)	3.3 (B)
Fumigants	0.3837	0.3 (A)	1.1 (A)
Organic synthesis compounds	0.9268	0.8 (A)	1.1 (A)

p-value - the attained significance level;
population-density populations that share the same letter (A,B) are not significantly different
* analyzed for four 1-km^2 grid cells adjacent to the drinking-water source
† significant at 95 percent confidence level
BTEX - benzene, toluene, ethylbenzene, and *m*-, *o*- and *p*-xylenes

The occurrence of MTBE in finished drinking water (Squillace and Moran 2000, Grady and Casey 2001) and in ambient ground- and surface waters (Moran et al. 1999a; Moran, et al. 1999b; Squillace et al. 1999; Moran et al. 2002b) also has been previously shown to be associated with patterns of MTBE use. Areas of the country that participate in the Federal RFG Program (U.S. Environmental Protection Agency 1999b) and (or) the winter OXY Program (U.S. Environmental Protection Agency 1999c), and where MTBE is known to be currently or formerly the oxygenate used to meet program requirements, have been found by the above investigators to correlate with higher MTBE occurrence. Similarly, the distribution of source waters that have concentrations of MTBE equal to or greater than the MRL also is related to the distribution of high MTBE-use areas (Figure 5.19). A statistical comparison of the frequency of detection of MTBE and other gasoline-related VOCs for areas that have and have not used fuels containing elevated concentrations of MTBE is summarized in Table 5.9.

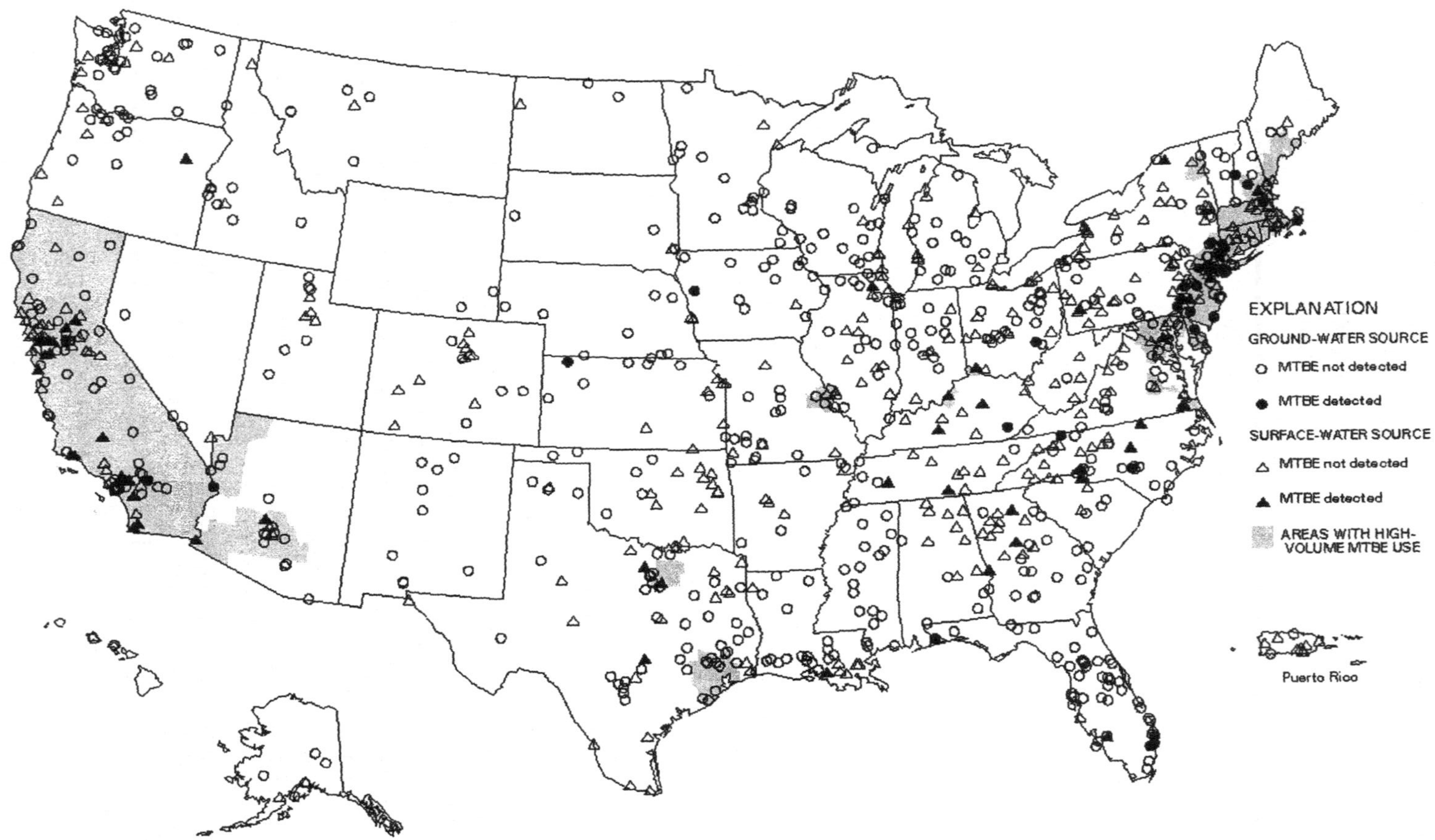

Figure 5.19. Distribution of MTBE detection in drinking-water sources in the United States and Puerto Rico in relation to high MTBE-use areas – Random Survey

MTBE was detected in drinking-water sources five times more frequently in the high MTBE-use areas than in areas that have not had comparable MTBE use. Although a similar pattern was observed for the subgroup "any gasoline compound," this is probably due to the fact that MTBE alone contributes two-thirds of all gasoline compound detections in source waters. BTEX compounds, which were comparatively rarely detected (1.9 percent) in source-water samples,were not statistically related to MTBE-use patterns. BTEX compounds, however, are intrinsic components of all types of gasoline and would not be expected to be statistically related to either high or low MTBE-use areas.

Table 5.9
Statistical comparison of the occurrence of MTBE and other gasoline-related volatile organic compounds in source-water samples by MTBE-use area – Random Survey

Volatile organic compound or related compounds	p-value	MTBE use* and frequency of detection (population)	
		Source waters not in high MTBE-use area	Source waters in high MTBE use area
Any gasoline compound	<0.0001†	6.2 (A)	23 (B)
Methyl *tert*-butyl ether	<0.0001†	4.4 (A)	23 (B)
BTEX compounds	0.9485	2.0 (A)	1.8 (A)

p-value - the attained significance level; values
* populations that share the same letter (A,B) are not significantly different
† significant at 95 percent confidence level
BTEX - benzene, toluene, ethylbenzene, and *m*-, *o*- and *p*-xylenes

Lastly, the occurrence of gasoline-related VOCs in source water was compared to the density of chemical storage tanks, the great majority of which stored gasoline near the drinking-water source. Information on the locations of above- and below-ground storage tanks, as well as information on the number of leaking underground storage tanks, was obtained from StarView Real Estate Version 2.6.1 (Vista Information Solutions 1999), and the density of these sites within the 4-km2 grid surrounding source-water locations was calculated. Drinking-water sources were then characterized as having storage-tank densities of 0, less than 1 (numerical average density), 1 to 5, 5 to 10, or more than 10 tanks/mi2 in their vicinity, and the frequency of detection of gasoline compounds was tested for independence with respect to tank-density category. The results of this analysis (Table 5.10) show that any gasoline compound, collectively, and BTEX compounds were detected significantly more often where there were more than 10 storage tanks per square mile. MTBE detections, conversely, were not related to storage-tank density at all. Although leaking underground storage tanks clearly have been sources of MTBE contamination of ground water (Happel, Dooher, and Beckenbach 1998; Hitzig, Kostecki, and Leonard 1998), other studies have similarly reported the lack of a statistically significant association between MTBE detections in drinking-water wells and the proximity (State of Maine 1998) or density (Shelton et al. 2001) of gasoline storage tanks. The lack of a statistically significant relation between MTBE detections and storage-tank density may reflect the enhanced mobility and recalcitrance of MTBE relative to most other gasoline compounds in ground water and (or) a greater variety of nonpoint sources such as small leaks and spills, urban runoff, recreational water-craft use, and atmospheric transport.

Table 5.10
Statistical comparison of the occurrence of MTBE and other gasoline-related volatile organic compounds in source-water samples by storage-tank density – Random Survey

Volatile organic compound or related compound	p-value	Storage-tank density* and frequency of detection (population)				
		0 tanks/mi^2	0 to 1 tanks/mi^2	1 to 5 tanks/mi^2	5 to 10 tanks/mi^2	More than 10 tanks/mi^2
Any gasoline compound	0.0356†	8.7 (A)	7.3 (A)	12 (AB)	15 (AB)	20 (B)
MTBE	0.4271	8.4 (A)	7.0 (A)	9.8 (A)	13 (A)	14 (A)
BTEX	0.0192†	1.3 (AB)	0.4 (A)	2.9 (AB)	3.6 (AB)	6.8 (B)

p-value - the attained significance level
storage-tank density populations that share the same letter (A,B) are not significantly different
* analyzed for four 1-km^2 grid cells adjacent to the drinking-water source
† significant at 95 percent confidence level
mi^2 - square mile
BTEX - benzene, toluene, ethylbenzene, and *m*-, *o*- and *p*-xylenes

FOCUSED SURVEY – MTBE AND OTHER VOCS

Implementation of the Focused Survey

Sample collection and analysis for the Focused Survey portion of the project began in September 1999 and was completed in April 2001. Samples were collected by USGS personnel, using USGS protocols, prior to any water treatment. Ground-water samples were collected from water-supply wells, as described by Koterba, Wilde, and Lapham (1995). Surface-water samples were collected as close to the intake as was reasonable and safe within 5 feet of the surface using a VOC hand-sampler as described by Shelton (1997) and Halde, Delzer, and Zogorski (1998).

A total of 451 samples were collected and analyzed from the 134 source waters selected—160 ground-water samples, 155 reservoir and (or) lake samples, and 136 river and (or) stream samples (Table 5.11). The scheduled number of samples were collected from most of the sites. However, 2 samples collected from reservoirs were lost in shipment, 5 (instead of 4) samples were collected from one reservoir, and 3 (instead of 2) samples were collected from three wells over the sampling period. The number of samples collected from systems suspected and known to have MTBE contamination are shown in Appendix I, Tables I.1 and I.2, respectively. The number of systems used as control sites are shown in Appendix I- Table I.3.

Table 5.11
Summary of number of source waters sampled, frequency of sample collection, and number of samples collected - Focused Survey

Source-water type	Number of CWS sources sampled	Frequency of Sample collection Number per year)	Total number of samples collected
Ground water	78	2	160
Reservoir/lake	39	4	155
River/stream	17	8	136
Total	134	--	451

CWS - community water system
-- not applicable

Review of the QC data indicates no systematic contamination of samples (Chapter 4). Detected VOC concentrations in blank samples typically were less than 0.2 µg/L, and very little variability was associated with the sample collection, transportation, and analytical processes for the majority of VOCs included in this study. A detailed review of these QC data is presented in Appendix G.

Occurrence and Temporal Variability of Volatile Organic Compounds

The results of the chemical analyses of source-water samples collected for the Focused Survey were evaluated for the occurrence and temporal variability of individual VOCs and subgroups of VOCs (Table 3.2 - Chapter 3). VOC subgroups examined include gasoline oxygenates, other gasoline related compounds (excluding the oxygenates), THMs, solvents, organic synthesis compounds, fumigants, and refrigerants. All source-water data used in these analyses have undergone an extensive QC screening procedure (Appendix G). This systematic procedure compares source-water data with respective FRB samples to ultimately accept or reject the source-water data. Thus, only those source-water data that met data-quality criteria were accepted and included in these analyses.

Concentrations of VOCs detected in the Focused Survey generally were less than 1 µg/L (Appendix I – Table I.4); however, the concentrations of benzene, tetrachloroethene, and trichloroethene in some samples exceeded the USEPA MCLs set for those compounds. It is important to note that MCLs are established for finished drinking water and not for source water. MCLs for benzene, tetrachloroethene, and trichloroethene have been set at 5 µg/L for each compound (U.S. Environmental Protection Agency 2001). Benzene concentrations exceeded the MCL in 4 samples, tetrachloroethene in 13 samples, and trichloroethene in 7 samples. Concentrations of MTBE detected in the Focused Survey exceeded the lower limit of the USEPA Drinking Water Advisory of 20 to 40 µg/L (U.S. Environmental Protection Agency 1997a) in 12 samples. All samples in which concentrations exceeded MCLs or the Drinking Water Advisory for MTBE were collected from ground-water source waters. Of the 4 control sites—2 wells, 1 reservoir, 1 river—that were selected because MTBE was not suspected to be detected, MTBE was detected only in 1 ground-water sample at a concentration of 0.05 µg/L.

Occurrence and Comparison to the Random Survey

One objective of the Random Survey was to provide a statistically representative assessment of the quality of the source water of the nation's public drinking-water supplies with respect to the occurrence of MTBE and other VOCs. This objective differs markedly from that of the Focused Survey, which intentionally sampled source waters suspected or known to be contaminated with MTBE. Some comparisons between the two surveys, however, are warranted.

Results from the Random Survey (Rowe et al. 2001; Grady, in press) were based on the analyses of source water obtained from 954 randomly selected CWSs sampled between May 3, 1999, and October 23, 2000. The Random Survey data underwent the same QC screening process that was applied to data from the Focused Survey. VOC concentrations detected during the Random Survey, however, were only reported at or above the MRL. As such, any meaningful comparisons between the two surveys had to be conducted with a common assessment level. Thus, the Focused Survey data were also censored at the MRL in order to make comparisons with the Random Survey.

Additional adjustments to Focused Survey data were needed in order to compare findings between the two surveys. Results from the Random Survey were commonly presented such that reservoir, lake, river, and stream source waters were combined into one surface-water category. Thus, all Focused Survey surface-water data were combined into one category. Finally, in the Random Survey one sample was collected from each CWS source water, whereas in the Focused Survey multiple samples were collected from the same CWS source water. Because concentrations may change from one sample to the next at the same site, one sample was randomly selected from each source water sampled as part of the Focused Survey and used for comparing results with the Random Survey.

As expected, most VOCs were detected more frequently in the Focused Survey than in the Random Survey (Figure 5.20). The Focused Survey sampled source waters suspected or known to contain MTBE whereas the Random Survey did not. Thus, MTBE was detected much more frequently at sites sampled in the Focused Survey than in the Random Survey. MTBE was detected in 60 and 49.1 percent of CWSs supplied by ground and surface water, respectively, in the Focused Survey and in 5.4 and 13.8 percent of CWSs supplied by ground and surface water, respectively, in the Random Survey. MTBE was detected in 55.5 percent of the 134 CWSs sampled in the Focused Survey and in 8.7 percent of 954 CWSs sampled in the Random Survey (Table 5.12).

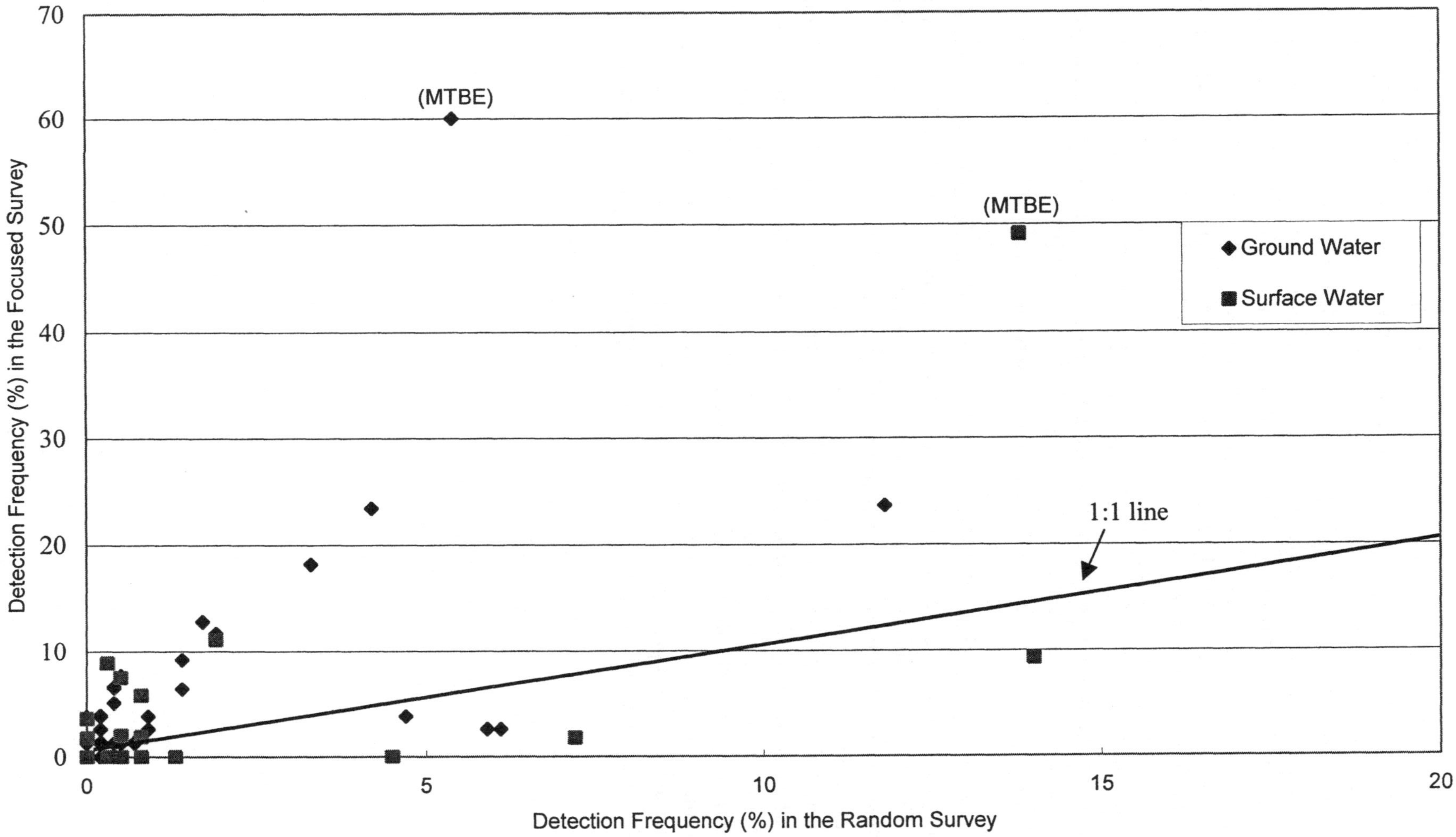

Figure 5.20 Detection frequency at or above the minimum reporting level of individual volatile organic compounds in source waters sampled in the Focused Survey versus the Random Survey

In addition to MTBE, numerous other VOCs were detected more frequently in the Focused Survey than in the Random Survey (Figure 5.20). This may be explained by population density in areas sampled in the Focused Survey. High-population density, which was one factor associated with the selection of a site to be sampled as part of the Focused Survey, has been shown to be related to increased VOC detection frequencies (Reiser and O'Brien 1998; Squillace et al. 1999; Bush et al. 2000; Lopes and Furlong 2001). Similarly, results from the Random Survey concluded that the occurrence of VOCs was greatest in the largest ground-water CWSs serving populations greater than 50,000 people and medium-sized surface-water CWSs serving populations of 3,001 to 10,000 people.

Individual VOC detection frequencies in CWSs sampled as part of the Focused and Random Surveys are presented in Table 5.12. Detection frequencies are presented for data at or above the MRL for both surveys and also at or above the MDL for the Focused Survey. Detection frequencies are lower when using data at or above the MRL because detections of VOCs at or above the MDL, yet below the MRL, were not included in the calculation (Appendix G).

When considering data above the MDL and MRL as part of the Focused Survey, about 93 and 61 percent of all source waters sampled contained at least one of the 66 VOCs analyzed, respectively. This compares to about 27 percent detected in source waters sampled as part of the Random Survey. Although detection frequencies were less in the Random Survey than in the Focused Survey, a similar number of VOCs was detected in both surveys. A total of 58 and 41 VOCs were detected at least one time in the Focused Survey above the MDL and MRL, respectively, whereas, a total of 42 were detected at levels above the MRL in the Random Survey. Thirty-seven of the 42 VOCs detected in the Focused Survey also were detected in the Random Survey. Again, only the detection frequencies above the MRL should be used for comparison to the Random Survey. A summary of detected concentrations of VOCs collected as part of the Focused Survey is presented in Appendix I – Table I.4. Additional discussion of the occurrence of VOCs in the Focused Survey above the MDL is included in the next section.

Table 5.12

Summary of detection frequencies of any volatile organic compound (VOC), any subgroup of VOCs, and individual VOCs in source waters sampled in the Focused and Random Surveys

Volatile organic compound subgroup and individual compound name	Focused Survey Detection frequency at or above the MDL (percent)			Detection frequency at or above the MRL (percent)			Random Survey Detection frequency at or above the MRL (percent)		
	All CWSs (n=134)	Ground Water (n=78)	Surface Water (n=56)	All CWSs (n=134)	Ground Water (n=78)	Surface Water (n=56)	All CWSs (n=954)	Ground Water (n=579)	Surface Water (n=375)
Any VOC	92.5	92.3	92.9	61.2	66.7	53.6	26.9	25.2	29.6
Gasoline Oxygenates									
tert-Amyl methyl ether	20.3	20.8	19.6	4.5	6.5	1.8	0.2	0.4	0.0
Diisopropyl ether	6.0	10.3	0.0	3.0	5.1	0.0	0.2	0.4	0.0
Ethyl *tert*-butyl ether	1.5	2.6	0.0	0.7	1.3	0.0	0.1	0.2	0.0
Methyl *tert*-butyl ether	77.3	68.0	90.6	55.5	60.0	49.1	8.7	5.4	13.8
Any gasoline oxygenate	73.9	65.4	85.7	53.7	59.0	46.4	8.5	5.3	13.3
Other Gasoline Compounds									
Benzene	22.0	15.8	30.4	7.6	6.6	8.9	0.3	0.4	0.3
n-Butylbenzene	1.5	1.3	1.8	0.0	0.0	0.0	0.1	0.2	0.0
sec-Butylbenzene	1.5	2.6	0.0	0.0	0.0	0.0	0.0	0.0	0.0
tert-Butylbenzene	0.0	0.0	0.0	0.0	0.0	0.0	0.0	0.0	0.0
Ethylbenzene	4.8	4.1	6.0	1.6	1.4	2.0	0.5	0.5	0.5
Naphthalene	9.9	10.5	9.1	2.3	3.9	0.0	0.2	0.2	0.3
Toluene	18.3	9.4	31.1	5.5	1.6	11.1	1.1	0.5	1.9
1,3,5-Trimethylbenzene	4.5	1.3	8.9	2.2	1.3	3.6	0.0	0.0	0.0
m,p-Xylene	8.6	5.3	13.5	3.1	1.3	5.8	0.8	0.7	0.8
o-Xylene	14.2	10.8	18.9	3.9	1.4	7.5	0.6	0.7	0.5
Any gasoline compound	35.8	29.5	44.6	9.0	7.7	10.7	2.2	1.9	2.7

(continued)

Table 5.12 (continued)

Volatile organic compound subgroup and individual compound name	Focused Survey Detection frequency at or above the MDL (percent) All CWSs (n=134)	Ground Water (n=78)	Surface Water (n=56)	Detection frequency at or above the MRL (percent) All CWSs (n=134)	Ground Water (n=78)	Surface Water (n=56)	Random Survey Detection frequency at or above the MRL (percent) All CWSs (n=954)	Ground Water (n=579)	Surface Water (n=375)
Trihalomethanes									
Bromoform	4.5	5.1	3.6	2.2	3.8	0.0	3.4	4.7	1.3
Bromodichloromethane	20.9	15.4	28.6	2.2	2.6	1.8	6.4	5.9	7.2
Chlorodibromomethane	12.7	10.3	16.1	1.5	2.6	0.0	5.5	6.1	4.5
Chloroform	48.4	55.6	38.9	17.5	23.6	9.3	12.7	11.8	14.0
Any trihalomethane	47.8	53.8	39.3	16.4	21.8	8.9	14.8	15.2	14.0
Solvents									
Bromobenzene	0.0	0.0	0.0	0.0	0.0	0.0	0.0	0.0	0.0
Carbon tetrachloride	2.2	3.8	0.0	1.5	2.6	0.0	0.7	0.9	0.5
Chlorobenzene	3.7	5.1	1.8	1.5	2.6	0.0	0.1	0.2	0.0
Chloroethane	0.7	1.3	0.0	0.0	0.0	0.0	0.1	0.2	0.0
2-Chlorotoluene	0.0	0.0	0.0	0.0	0.0	0.0	0.0	0.0	0.0
4-Chlorotoluene	0.7	1.3	0.0	0.0	0.0	0.0	0.0	0.0	0.0
Dibromomethane	0.7	1.3	0.0	0.7	1.3	0.0	0.2	0.4	0.0
1,2-Dichlorobenzene	1.5	2.6	0.0	0.7	1.3	0.0	0.1	0.2	0.0
1,3-Dichlorobenzene	2.2	3.8	0.0	0.0	0.0	0.0	0.0	0.0	0.0
1,1-Dichloroethane	15.9	26.3	1.8	5.3	9.2	0.0	1.2	1.4	0.0
1,2-Dichloroethane	11.9	20.5	0.0	4.5	7.7	0.0	0.3	0.5	0.0
1,1-Dichloroethene	6.0	10.3	0.0	3.7	6.4	0.0	0.8	1.4	0.0
cis-1,2-Dichloroethene	12.8	22.1	0.0	6.8	11.7	0.0	1.3	1.9	0.3
trans-1,2-Dichloroethene	4.5	7.7	0.0	2.2	3.8	0.0	0.0	0.0	0.0
1,2-Dichloropropane	3.7	6.4	0.0	0.7	1.3	0.0	0.1	0.2	0.0
Hexachlororethane	0.0	0.0	0.0	0.0	0.0	0.0	0.0	0.0	0.0
Methylene chloride	0.8	1.4	0.0	0.8	1.4	0.0	0.2	0.2	0.3
Methyl ethyl ketone	0.8	1.3	0.0	0.0	0.0	0.0	0.5	0.4	0.8

(continued)

Table 5.12 (continued)

Volatile organic Compound subgroup and individual compound name	Focused Survey Detection frequency at or above the MDL (percent)			Detection frequency at or above the MRL (percent)			Random Survey Detection frequency at or above the MRL (percent)		
	All CWSs (n=134)	Ground Water (n=78)	Surface Water (n=56)	All CWSs (n=134)	Ground Water (n=78)	Surface Water (n=56)	All CWSs (n=954)	Ground Water (n=579)	Surface Water (n=375)
Solvents (cont.)									
n-Propylbenzene	3.0	1.3	5.4	0.8	1.3	0.0	0.1	0.0	0.3
1,1,1,2-Tetrachloroethane	1.5	2.6	0.0	0.0	0.0	0.0	0.0	0.0	0.0
Tetrachloroethene	21.1	36.4	0.0	13.5	23.4	0.0	2.8	4.2	0.8
1,1,2,2-Tetrachloroethane	0.0	0.0	0.0	0.0	0.0	0.0	0.0	0.0	0.0
1,2,4-Trichlorobenzene	3.0	2.6	3.6	0.0	0.0	0.0	0.0	0.0	0.0
1,1,1-Trichloroethane	14.2	24.4	0.0	7.5	12.8	0.0	1.0	1.7	0.0
1,1,2-Trichloroethane	1.5	2.6	0.0	0.0	0.0	0.0	0.0	0.0	0.0
Trichloroethene	17.3	29.9	0.0	10.5	18.2	0.0	2.3	3.3	0.8
1,2,3-Trichloropropane	3.8	3.8	3.7	0.0	0.0	0.0	0.1	0.2	0.0
Any solvent	41.8	61.5	14.3	21.6	37.2	0.0	6.3	8.8	2.4
Organic Synthesis									
Acrylonitrile	0.7	1.3	0.0	0.7	1.3	0.0	0.0	0.0	0.0
Bromochloromethane	3.7	5.1	1.8	0.0	0.0	0.0	0.0	0.0	0.0
1,3-Dichloropropane	3.0	2.6	3.6	0.7	1.3	0.0	0.0	0.0	0.0
2,2-Dichloropropane	0.0	0.0	0.0	0.0	0.0	0.0	0.0	0.0	0.0
1,1-Dichloropropene	0.0	0.0	0.0	0.0	0.0	0.0	0.0	0.0	0.0
Hexachlorobutadiene	0.0	0.0	0.0	0.0	0.0	0.0	0.0	0.0	0.0
Isopropylbenzene	3.8	5.2	1.8	0.8	1.3	0.0	0.2	0.2	0.3
p-Isopropyltoluene	0.7	0.0	1.8	0.0	0.0	0.0	0.0	0.0	0.0
Styrene	16.3	21.3	9.3	3.8	6.6	0.0	0.5	0.4	0.8
1,2,4-Trimethylbenzene	7.6	1.3	16.7	1.5	1.3	1.9	0.4	0.2	0.8
1,2,3-Trichlorobenzene	5.2	6.4	3.6	0.0	0.0	0.0	0.0	0.0	0.0
Vinyl bromide	0.0	0.0	0.0	0.0	0.0	0.0	0.0	0.0	0.0
Vinyl chloride	2.2	3.8	0.0	2.2	3.8	0.0	0.1	0.2	0.0
Any organic synthesis	32.8	33.3	32.1	7.5	11.5	1.8	1.0	0.7	1.6

(continued)

Table 5.12 (continued)

Volatile organic compound subgroup and individual compound name	Focused Survey Detection frequency at or above the MDL (percent) All CWSs (n=134)	Ground Water (n=78)	Surface Water (n=56)	Detection frequency at or above the MRL (percent) All CWSs (n=134)	Ground Water (n=78)	Surface Water (n=56)	Random Survey Detection frequency at or above the MRL (percent) All CWSs (n=954)	Ground Water (n=579)	Surface Water (n=375)
Fumigants									
Bromomethane	3.0	3.8	1.8	0.0	0.0	0.0	0.2	0.2	0.3
1,4-Dichlorobenzene	4.5	6.5	1.8	0.8	1.3	0.0	0.3	0.2	0.5
cis-1,3-Dichloropropene	0.0	0.0	0.0	0.0	0.0	0.0	0.0	0.0	0.0
trans-1,3-Dichloropropene	0.0	0.0	0.0	0.0	0.0	0.0	0.0	0.0	0.0
Any fumigant	7.5	10.3	3.6	0.7	1.3	0.0	0.5	0.4	0.8
Refrigerants									
Chloromethane	3.0	3.8	1.8	0.7	1.3	0.0	0.3	0.5	0.0
Dichlorodifluoromethane	0.7	1.3	0.0	0.7	1.3	0.0	0.4	0.7	0.0
1,1,2-Trichloro-1,2,2-triflu-	0.7	1.3	0.0	0.7	1.3	0.0	0.1	0.2	0.0
Trichlorofluoromethane	2.2	3.8	0.0	2.2	3.8	0.0	0.5	0.9	0.0
Any refrigerant	5.2	7.7	1.8	3.0	5.1	0.0	1.0	1.7	0.0

VOC - volatile organic compound
CWS - community water system
n - number of samples]

Each subgroup of VOCs was detected at or above the MRL at a higher frequency in the Focused Survey than in the Random Survey (Table 5.12, Figure 5.21). The largest difference in detection frequencies was found in the gasoline oxygenate subgroup, wherein these compounds were detected in 53.7 percent of CWSs sampled in the Focused Survey but in only 8.5 percent of CWSs sampled in the Random Survey. This difference is due to the intentional selection of sites with suspected or known occurrence of MTBE in the Focused Survey. A slightly higher detection frequency for gasoline oxygenates was found in ground-water source waters (59.0 percent) than in surface-water source waters (46.4 percent) in the Focused survey. This pattern was reversed, however, for samples collected in the Random Survey, where gasoline oxygenates were detected in 5.3 and 13.3 percent of ground and surface water, respectively. MTBE was the most frequently detected gasoline oxygenate in source waters sampled in the Focused and Random Surveys, occurring in 55.5 and 8.7 percent, respectively.

A second large difference was found in the solvent subgroup, again with the larger detection frequency occurring in the Focused Survey. In both surveys, most of the solvent detections were in ground-water source waters (Table 5.12). This may be due to solvents being removed or diluted slower in ground water than in surface water. The difference in detection frequency between the two surveys (Figure 5.21) may be due, in part, to intentionally sampling ground-water sites as part of the Focused Survey that were considered more vulnerable than other sites. That is, preference was given to sampling ground-water source waters withdrawn from shallow, unconsolidated, unconfined aquifers in urban areas serving large populations. The solvents detected most frequently in both surveys were tetrachloroethene and trichloroethene.

Interestingly, THMs were detected in the Focused and Random Surveys at similar frequencies—16.4 and 14.8 percent of CWSs, respectively. Chloroform was the most frequently detected THM in both surveys. However, chloroform was detected more frequently in ground-water source waters in the Focused Survey (23.6 percent) than in the Random Survey (11.8 percent). Similar THM detection frequencies in both surveys indicates a pervasive occurrence of THMs in several source-water types regardless of CWS size or location.

Other gasoline compounds, organic synthesis compounds, fumigants, and refrigerants all were detected at or above the MRL in less than 10 percent of CWSs sampled in each survey. Although the gasoline oxygenate MTBE was detected frequently in the Focused Survey, VOCs in the other gasoline compound subgroup were detected in only 9 percent of CWS source waters. This may be due to these compounds degrading and (or) volatilizing more rapidly in ground and surface water in comparison to MTBE. Organic synthesis compounds also were detected more frequently in the Focused Survey (7.5 percent) than in the Random Survey (1.0 percent). The majority of detections in the Focused Survey was found in ground-water source waters (11.5 percent) rather than in surface-water source waters (1.8 percent). Again, this may be due to the Focused Survey's sampling urban ground-water source waters in shallow, unconsolidated, unconfined aquifers that may be susceptible to contamination. Very little difference in detection frequencies between either survey was observed for fumigants and refrigerants.

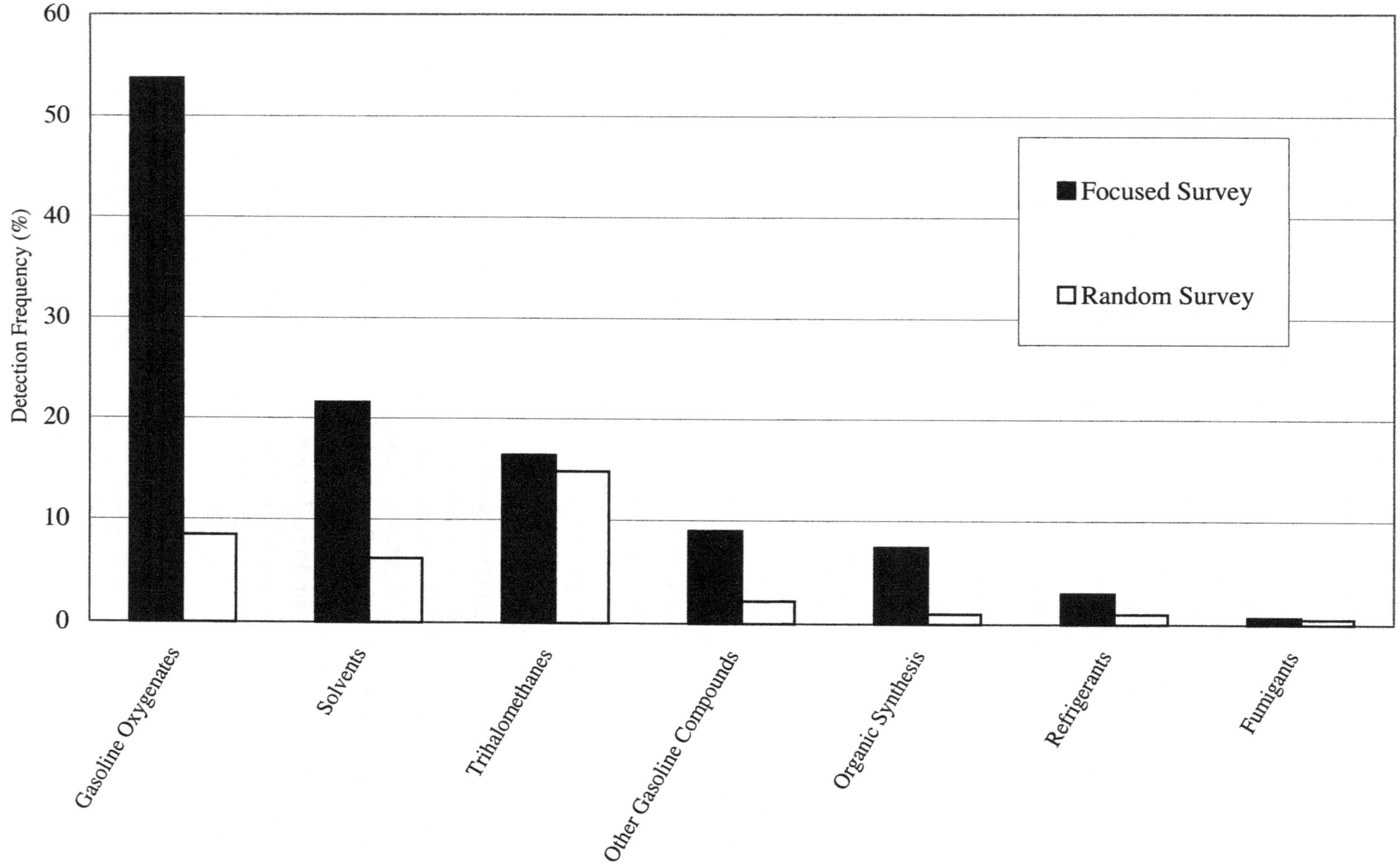

Figure 5.21 Detection frequency at or above the minimum reporting limit of volatile organic compound subgroups in community water system source waters sampled in the Focused versus the Random Surveys

CWSs were subdivided into 5 size categories based on population served: (1) very small (serves 25 to 500 people); (2) small (serves 501 to 3,300 people); (3) medium (serves 3,301 to 10,000 people); (4) large (serves 10,001 to 50,000 people); and (5) very large (serves more than 50,000 people). Table 5.13 lists the relative percent of CWSs per size of system sampled in the Focused and Random Surveys. A summary of source waters sampled per state by size of CWS is presented in Appendix I – Tables I.1 and I.2.

A similar distribution of ground-water CWS sizes were sampled in the Focused Survey—from 10.4 to 26 percent of each size category—whereas the majority of ground-water source waters sampled (50.4 percent) in the Random Survey were very small systems. The abundance of very small ground-water source waters in the Random Survey is due to the relative abundance of very small ground water supplied CWSs on a national basis and reflects the random selection process used in the Random Survey. A disproportionate number of very large CWSs supplied by surface water were sampled (85.7 percent) in the Focused Survey. Similarly, the majority of CWSs supplied by surface water sampled (81.9 percent) as part of the Random Survey were large or very large CWSs. The abundance of large and very large surface-water systems sampled as part of the Random Survey was due to the cumulative population served, which was one weighting factor used in the design of that survey (Chapter 3).

Table 5.13
Distribution of source waters sampled in the Focused and Random Surveys by source-type and size of community water system based on population served

Community water system size (based on population served)	System	
	Focused	Random
Percent of ground-water systems sampled		
Very small	10.4	50.4
Small	16.8	20.9
Medium	20.8	8.6
Large	26.0	10.9
Very large	26.0	9.2
Total	100.0	100.0
Percent of surface-water systems sampled		
Very small	0.0	3.2
Small	7.1	6.9
Medium	1.8	8.0
Large	5.4	21.0
Very large	85.7	60.9
Total	100.0	100.0

The highest VOC detection frequency in the Random Survey for each subgroup of VOCs (except THMs) was found in the very large CWSs supplied by ground water (Figure 5.22). This likely reflects the fact that many more of the larger systems have source waters within high-population density, urban areas than do the smaller CWSs. This pattern was not evident in the Focused Survey for gasoline oxygenates, other gasoline compounds, solvents, and refrigerants (Figure 5.22). This is likely due to the fact that the source waters selected were considered more vulnerable to VOC contamination and MTBE was suspected or known to occur in the source, regardless of CWS size.

In the Random Survey, gasoline oxygenates and other gasoline compounds were detected more frequently in the very large CWSs supplied by surface water than in other size categories (Figure 5.23). This pattern was generally observed in the Focused Survey. Although the Focused and Random Surveys each sampled disproportionately more large and very large CWSs than in other size categories, gasoline oxygenates, other gasoline compounds, solvents, THMs, and organic synthesis compounds were detected in virtually every size category sampled in the Random Survey. This was not observed in the Focused Survey. Gasoline oxygenates primarily were detected in large and very large CWSs; however, solvents, THMs, and organic synthesis compounds were not detected at all or only detected in a few sizes of CWSs. The reason for this difference is not known.

Temporal Variability

Samples were collected for one year at fixed intervals during the Focused Survey. Ground-water source waters were sampled biannually, reservoir and lake source waters were sampled quarterly, and river and stream source waters were sampled eight times per year (Table 5.11). The intent of the fixed-interval sampling was to obtain sufficient data to evaluate temporal variability of VOCs in each source-water type. To do so, samples were collected more frequently in surface-water source waters than in ground-water source waters to account for likely changes in water quality caused by temperature and other seasonal affects.

The ground-water sampling design did not specifically distribute samples seasonally as was done for surface-water sampling. Ground-water samples were collected approximately six months apart and only two samples were collected from each well. Because of the slow movement of ground water, the elapsed time between samples was presumed insufficient to allow for significant changes in ground-water quality to occur. Thus, analyses of temporal variability in ground-water source waters compare only the first sample to the second sample and do not evaluate seasonal patterns. In contrast, the analyses of temporal variability in surface-water source waters do evaluate seasonal patterns.

The four seasons considered in the temporal analyses were defined as spring (April - June), summer (July - September), fall (October - December), and winter (January - March). The number of reservoir and lake source-water samples collected were distributed fairly evenly among the different seasons (Figure 5.24). The number of reservoir/lake samples collected ranged from 37 during winter to 46 in the spring. However, more samples were collected from river and stream source waters during winter than in other seasons; about 30 river/stream samples were collected during the spring, summer, and fall, and 52 samples were collected during the winter. Generally, this may result in an increase in VOC occurrence in rivers and streams during the winter months due to higher air-to-water partitioning and decreased volatility of VOCs at winter temperatures.

Focused Survey

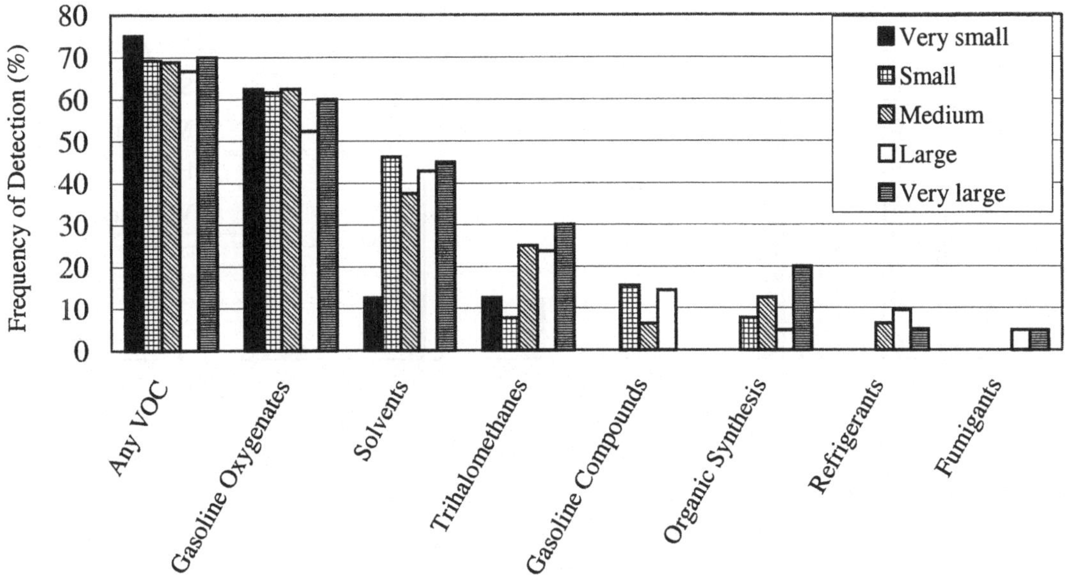

Random Survey

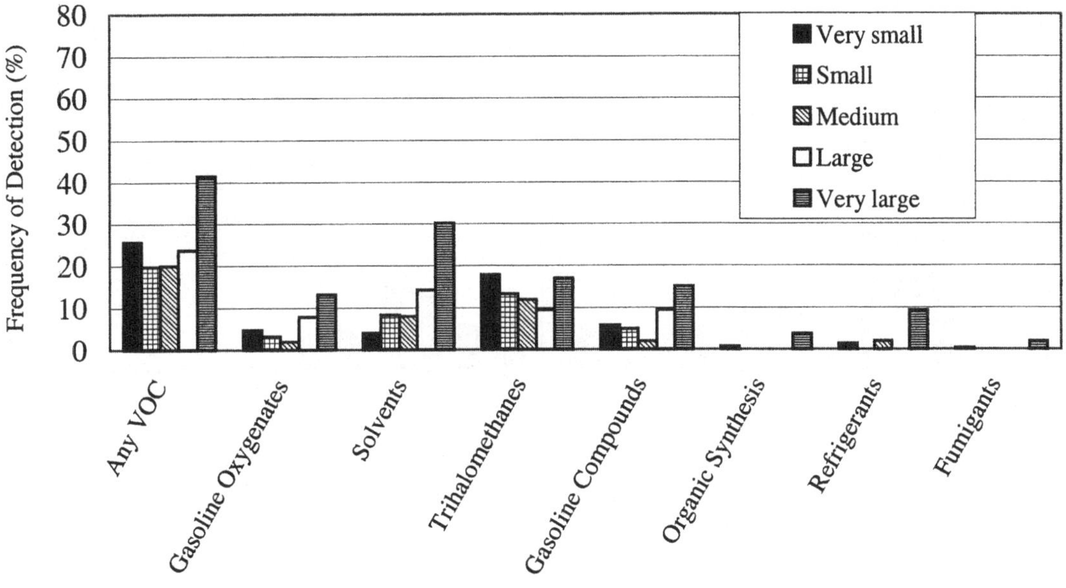

Figure 5.22 Detection frequency at or above the MRL of any VOC and subgroup in ground-water sources sampled in the Focused and Random Surveys by size of community water system

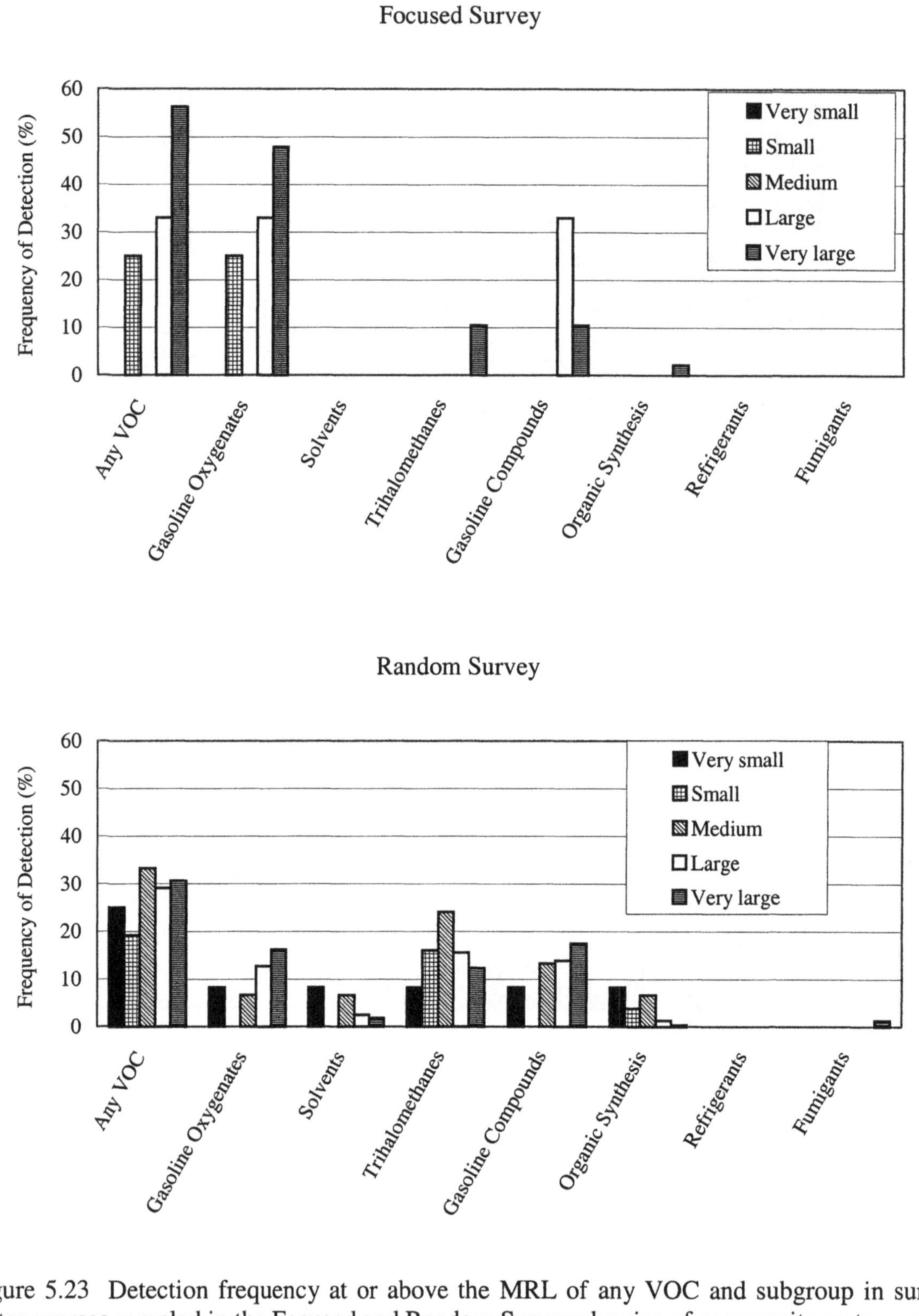

Figure 5.23 Detection frequency at or above the MRL of any VOC and subgroup in surface-water sources sampled in the Focused and Random Surveys by size of community water system

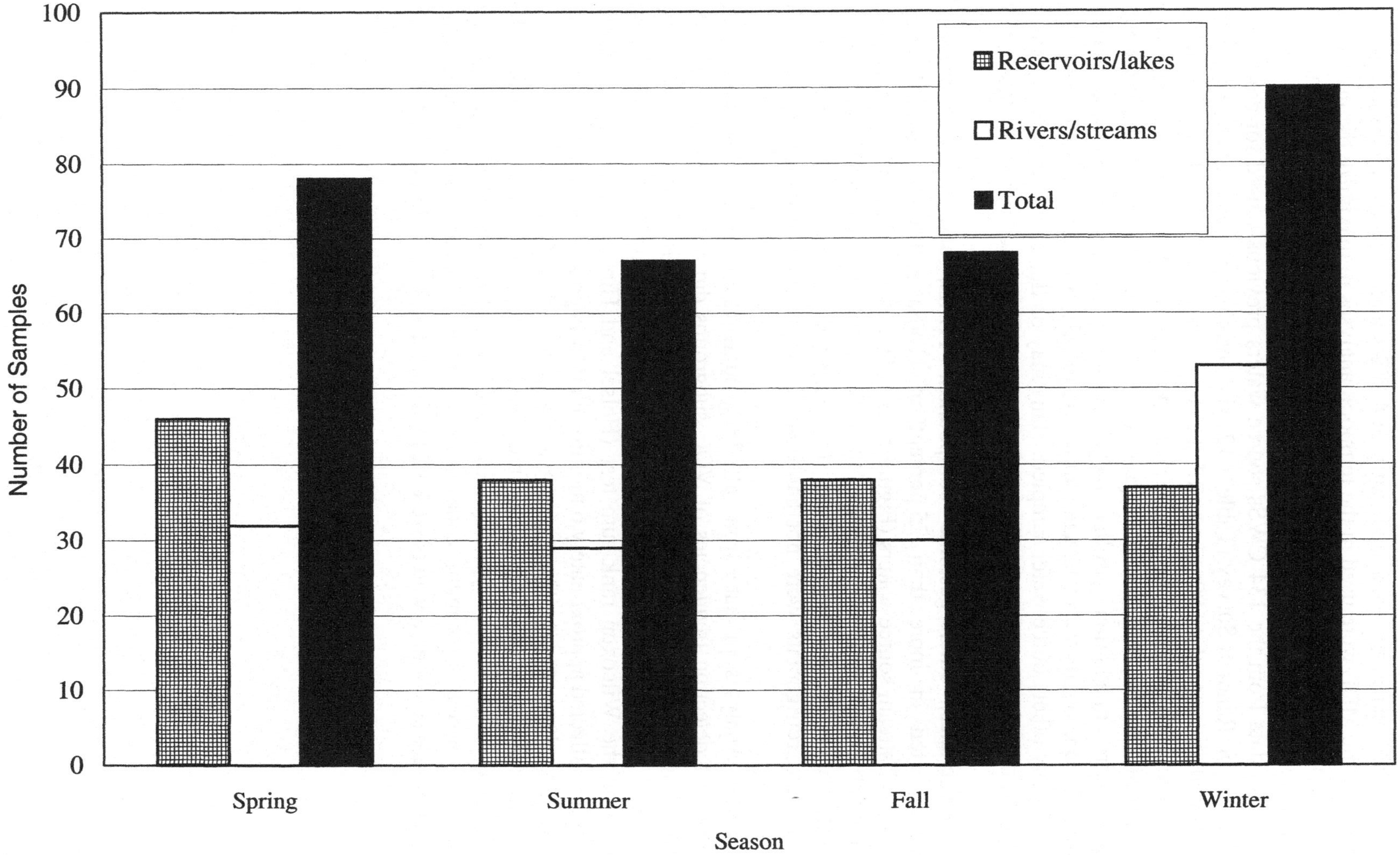

Figure 5.24 Number of surface-water samples collected by source type and season - Focused Survey

Temporal analyses were performed to examine seasonal and concentration differences in those subgroups of VOCs that occur most frequently in CWS source water. It is important to note that all data above the MDL were used in the temporal analyses to maximize the number of VOC detections that are available for analyses. A summary of concentrations detected in source water is presented in Appendix I – Table I.4.

Analyses of the data from the 134 CWSs source waters previously used for comparing the Focused Survey to the Random Survey (Table 5.12) showed that at least 1 of the 66 VOCs analyzed were detected in more than 90 percent of the source waters sampled as part of the Focused Survey (Figure 5.25). Each of the 17 CWSs supplied by rivers or streams contained at least one VOC, a result of the detection of a gasoline oxygenate in each of the 17 CWS river/stream source waters. In addition, gasoline oxygenates, other gasoline compounds, THMs, and organic synthesis compounds were detected more frequently in CWSs supplied by river source waters than in any other source water sampled. This may be due to the presence of point sources upstream from the sampling site that provide a fairly continuous source of contamination.

Gasoline oxygenates, other gasoline compounds, THMs, and organic synthesis compounds were detected in more than 25 percent of CWSs served by ground-water, reservoir/lake, and river/stream source waters (Figure 5.25). Therefore, these four subgroups of VOCs were used in subsequent temporal analyses. In addition, solvents were evaluated only in ground-water source waters because they were not detected frequently in surface-water source waters. Fumigants and refrigerants were not analyzed for temporal variability due to their infrequent detection.

Ground water. Analyses of data from all 78 ground-water source waters sampled biannually revealed that detection frequencies of VOC subgroups did not change significantly from the first sample collected to the second sample (Figure 5.26). Similarly, no statistically significant differences in detected concentrations of VOCs were found at the 95 percent confidence level using the Wilcoxon rank sum test (Helsel and Hirsch 1992) (Figure 5.27). Because samples were collected approximately 6 months apart, it is presumed that an insufficient amount of time passed to show any significant changes in ground-water quality.

Although these subgroups of VOCs occur frequently in the source waters sampled, the concentrations detected were small, typically less than 1 µg/L. However, some higher concentrations were observed—maximum concentrations of MTBE, tetrachloroethene, and trichloroethene in ground-water source waters were 924, 290, and 165 µg/L, respectively.

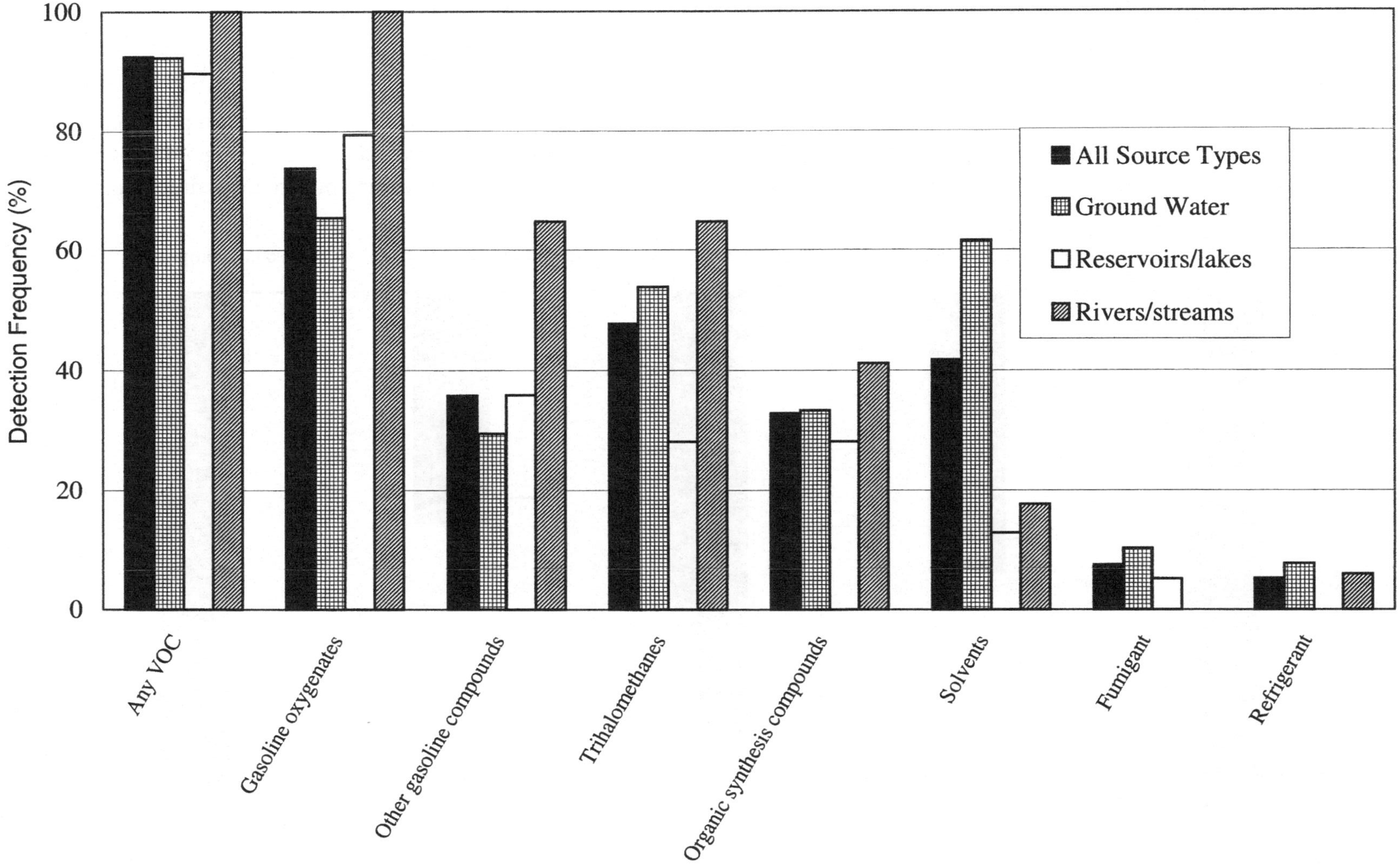

Figure 5.25 Detection frequency at or above the minimum reporting level of any volatile organic compound and by subgroup of VOCs in community water system source waters by source type - Focused Survey

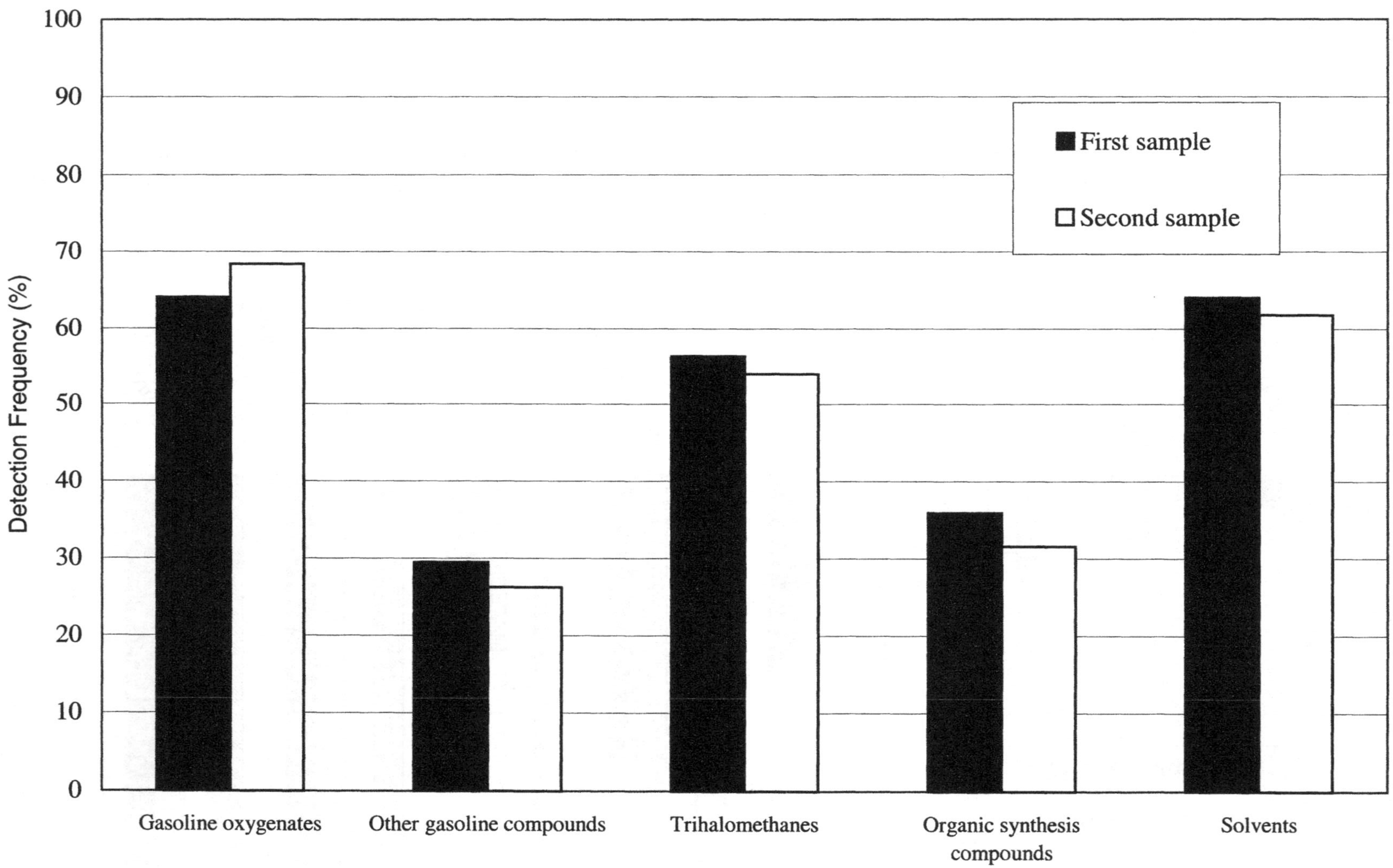

Figure 5.26 Comparison of the detection frequency of at least one volatile organic compound (VOC) at or above the method detection limit by subgroup of VOCs in the first and second sampling of ground-water sources - Focused Survey

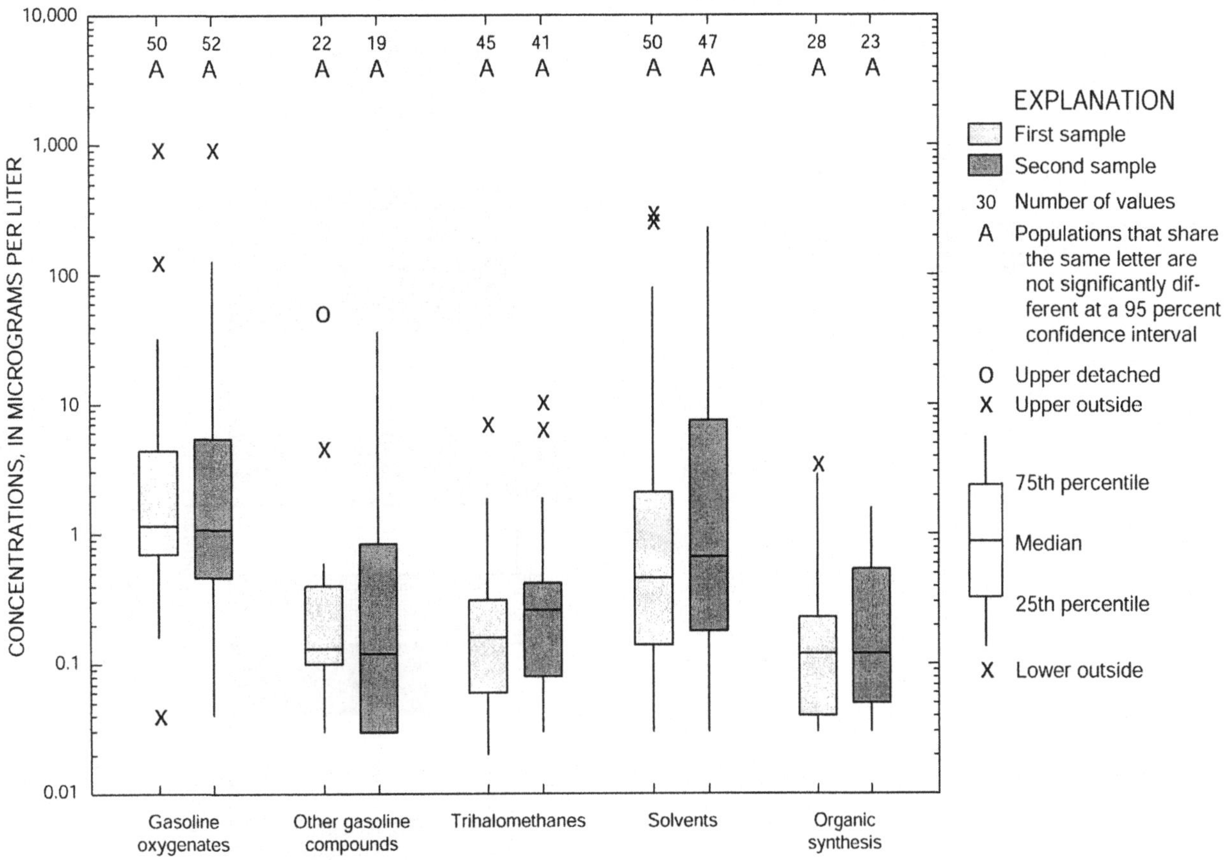

Figure 5.27 Detected concentrations of volatile organic compound subgroups in ground-water source water collected in the first and second samples – Focused Survey

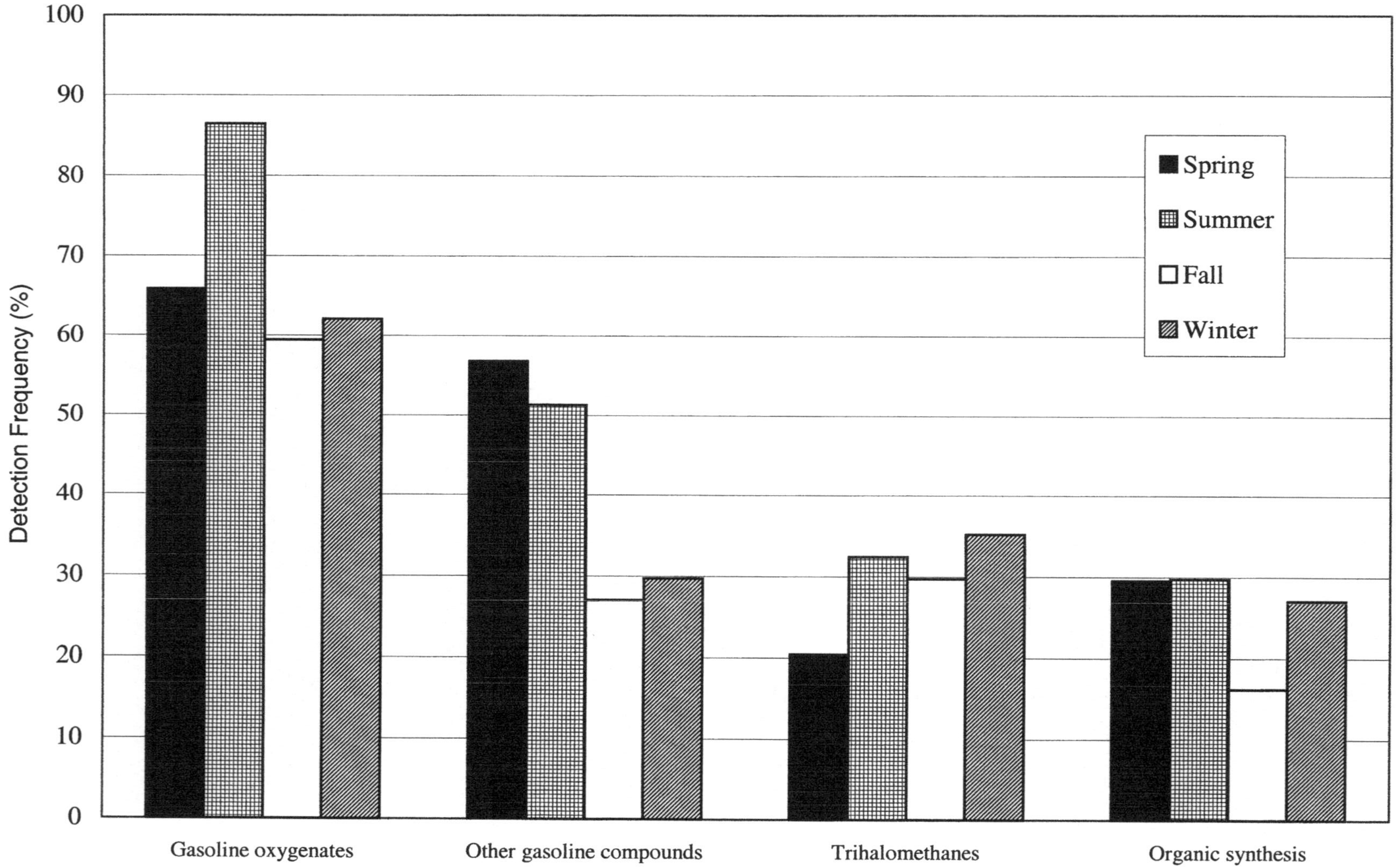

Figure 5.28 Detection frequency of at least one volatile organic compound (VOC) at or above the method detection limit by subgroup of VOCs in reservoir and lake sources by season -Focused Survey

Reservoirs and lakes. A slight seasonal pattern was evident in the occurrence of gasoline oxygenates and other gasoline compounds (Figure 5.28) in reservoir and lake source waters. Analyses of data from all 39 reservoirs and (or) lakes sampled quarterly revealed that both of these subgroups of VOCs were detected more frequently in the spring and summer than in the fall and winter. However, these compounds were still fairly prevalent during the fall and winter, when they were detected in at least 27 percent of samples. Thirty-three of the 39 reservoirs sampled allow motorized watercraft. Thus, the greater detection of gasoline related compounds in the spring and summer is presumed to be related to the increased use of motorized watercraft during these seasons. The continued detection of these compounds during fall and winter may be due, in part, to the continued use of watercraft containing reformulated gasoline during these seasons in areas with warmer climates, and the characteristics of MTBE which allow it to persist longer in water in comparison to other VOCs.

Trihalomethanes and organic synthesis compounds were detected in about 30 percent of samples collected throughout the year with no significant trend identified. The lack of any substantial, seasonal variability in THMs and organic synthesis compounds indicates the possibility of common and continuous source of contamination for these compounds.

Concentrations of VOCs detected in reservoir and lake source waters were typically less than 1 μg/L and very few statistically significant differences in seasonal concentrations were identified (Figure 5.29). Although gasoline oxygenates and other gasoline compounds were detected more frequently in the spring and summer than in fall and winter, statistically significant differences in detected median concentrations of VOCs typically were not identified by the Wilcoxon rank sum test at an alpha of 0.05. There were no statistically significant differences between seasonal concentrations of gasoline oxygenates and THMs. Statistically significant differences were found for other gasoline compounds and organic synthesis compounds. For these VOC subgroups, winter and (or) fall concentrations were significantly less than spring and (or) summer concentrations.

Rivers and streams. No seasonal pattern was evident for any of the more frequently occurring subgroups of VOCs in river and stream source waters (Figure 5.30). The detection of gasoline oxygenates was found to drop slightly from spring to winter; however, these VOCs were found to be consistently present in the source waters sampled—in at least 78 percent of the samples. This may be indicative of a common source of gasoline oxygenates to the source waters sampled. Other gasoline compounds, THMs, and organic synthesis compounds also were detected at similar frequencies throughout the year. Other gasoline compounds and THMs were detected in more than 50 percent of all samples and organic synthesis compounds were detected in more than 20 percent of all samples collected throughout the year. Again, the absence of seasonal differences for these subgroups of compounds may indicate an unknown, but common and continuous source of contamination.

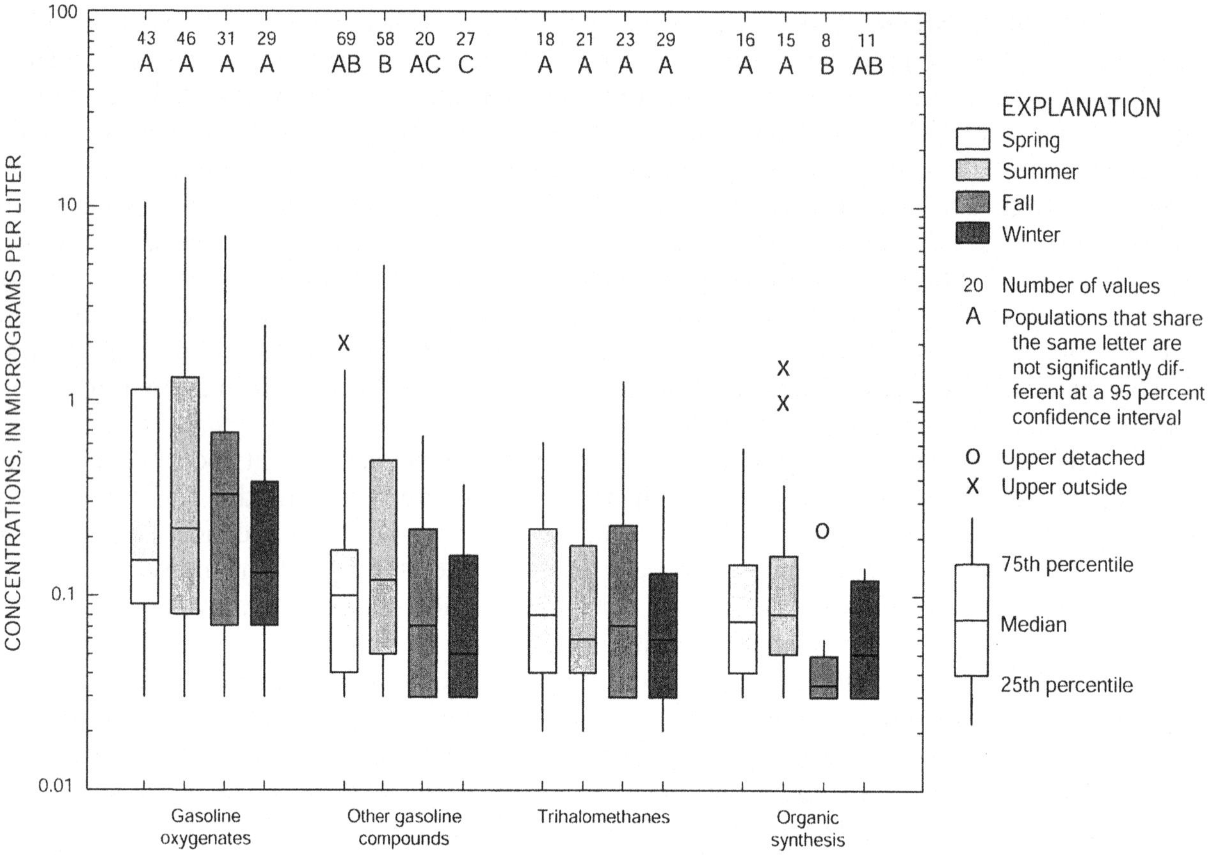

Figure 5.29 Detected concentrations of volatile organic compound subgroups in reservoir and lake source waters by season – Focused Survey

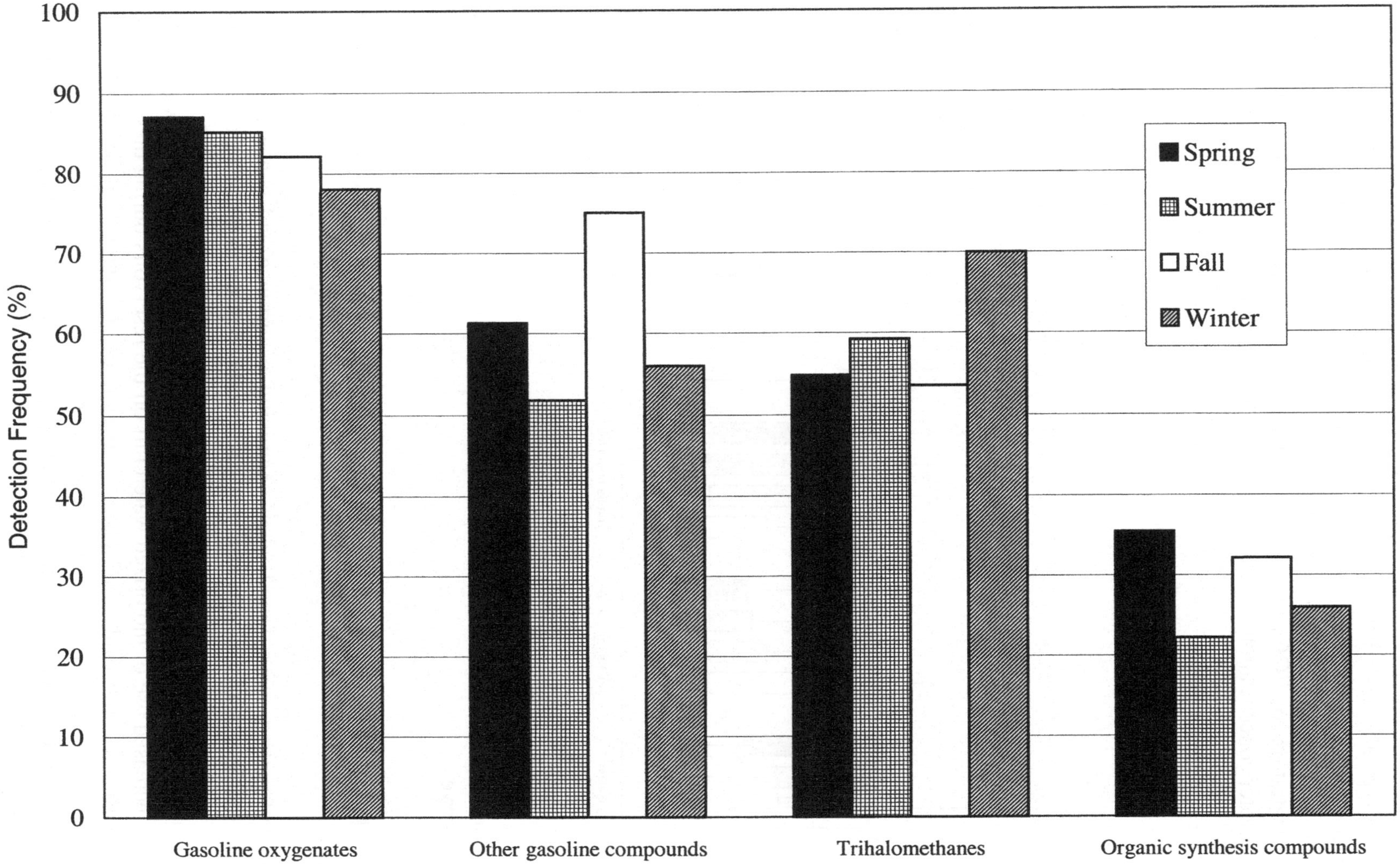

Figure 5.30 Detection frequency of at least one volatile organic compound (VOC) at or above the method detection limit by subgroup of VOCs in river and stream sources by season -Focused Survey

Concentrations of the VOCs (subgroups) most frequently detected in river and stream source waters were typically less than 1 μg/L, and few statistically significant differences in concentrations were noticed on a seasonal basis (Figure 5.31). Although these subgroups of VOCs were found frequently year round (Figure 5.30), the only statistically significant difference in the concentrations of VOCs were within the gasoline oxygenate subgroup, which were found at slightly greater concentrations during the summer than during the winter season.

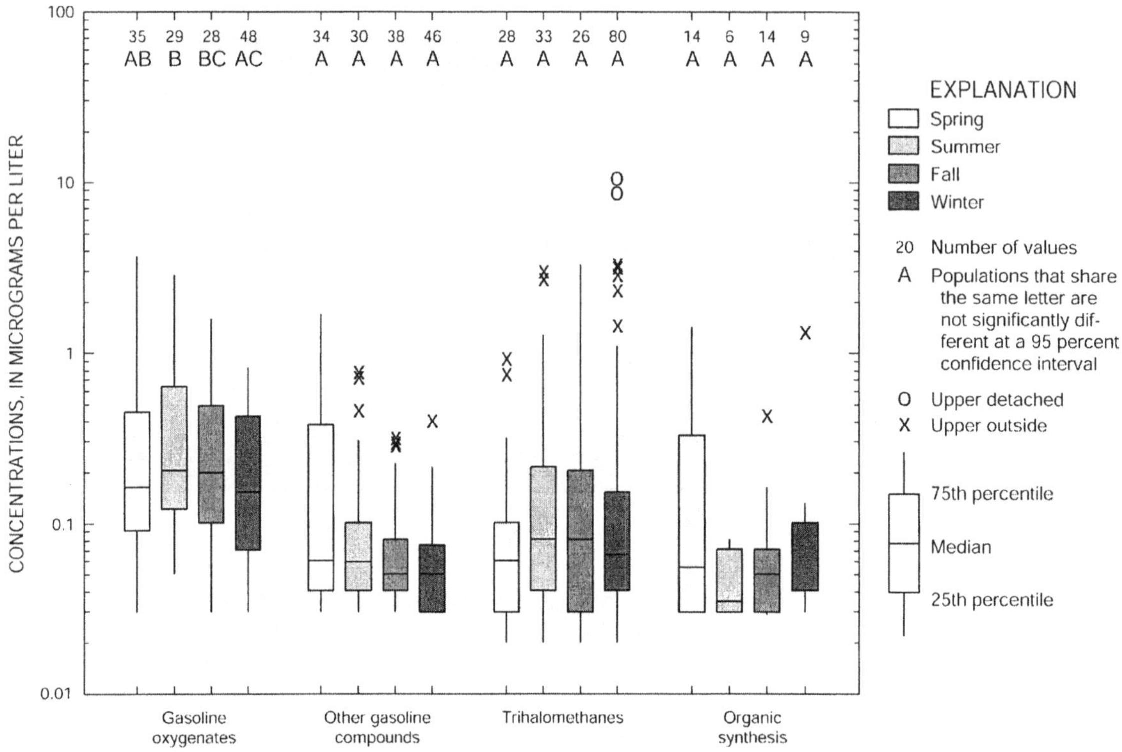

Figure 5.31 Detected concentrations of VOC subgroups in river and stream source waters by season – Focused Survey

Anthropogenic Factors Associated with MTBE, Gasoline Oxygenates, and Trihalomethanes

Few temporal patterns were found when examining occurrence and concentrations of VOC subgroups in source waters where MTBE was suspected or known to occur. One apparent trend was found in reservoirs and lakes, where gasoline oxygenates and other gasoline compounds were detected more frequently during the spring and summer. This was likely associated with the increased use of motorized watercraft on the reservoirs during these seasons. However, no significant statistical differences were noticed when examining MTBE concentrations alone on a seasonal basis. Although one or two VOCs typically dominate the occurrence within VOC subgroups (Table 5.13), the increased number of potential detections in subgroups should be beneficial in detecting significant trends not otherwise noticed when examining individual VOCs. As such, the most frequently detected VOC (MTBE) and the two most frequently occurring subgroups of VOCs (gasoline oxygenates and THMs) in the Focused Survey were further characterized to evaluate anthropogenic factors that may better explain and (or) to understand their frequent occurrence.

The two most frequently occurring subgroups of VOCs in the Focused Survey were gasoline oxygenates and THMs; these subgroups were detected in 73.9 and 47.8 percent of CWS source waters sampled, respectively (Table 5.13). In CWS source waters, MTBE was the most frequently detected gasoline oxygenate (77.3 percent) followed by TAME (20.3 percent). Chloroform and bromodichloromethane were the two most frequently detected THMs, occurring in 48.4 and 20.9 percent of CWS source waters, respectively.

Anthropogenic factors associated with gasoline oxygenates were evaluated on the basis of high-MTBE use areas for CWSs served by ground-water, reservoir and (or) lake, and river and (or) stream source waters, and on watercraft use on reservoirs and lakes. Anthropogenic factors associated with THMs detected in reservoir source waters were evaluated on the basis of ancillary data provided by CWS personnel. All analyses were performed using data at or above the MDL.

MTBE and Gasoline Oxygenates

Detected concentrations of MTBE in ground-water source waters inside and outside of high-MTBE use areas were compared. Fifty-five wells were within high-MTBE use areas and 23 were outside of these areas. MTBE was detected more frequently in low-MTBE use areas than in high-MTBE use areas. When both samples collected from individual wells were considered, MTBE was detected in 73.9 percent (34 of 46) of ground-water samples collected in low-MTBE use areas and in 62.7 percent (69 of 110) of ground-water samples collected in high-MTBE use areas. The differences in occurrence and concentration may be explained, in part, by how these sites were selected. Thirty-eight CWS wells were selected because of the suspected occurrence of MTBE and 37 were selected because of a known occurrence (Appendix I – Table I.1). However, 21 of the 37 sites where MTBE was known to occur were not in areas of high-MTBE use. This comparison is then skewed toward those sites outside of high-MTBE use areas and is merely a result of intentionally selecting wells with a known occurrence of MTBE. For this reason, statistical comparisons were not performed.

Concentrations of MTBE in reservoir and (or) lake source waters inside and outside of high-MTBE use areas were compared seasonally (Figure 5.32). Twenty-five reservoir/lake sources were inside high-MTBE use areas and 14 were outside of these areas. MTBE was detected more frequently and concentrations were greater in high-MTBE use areas than in low-MTBE use areas. This is likely due, in part, to the much larger percentage of MTBE volume/volume present in gasoline in high-MTBE use areas than in other areas. However, because as few as 4 detections of MTBE were found in reservoir/lake source waters in low-MTBE areas, statistical comparisons were not completed.

Concentrations of MTBE in river and (or) stream source waters inside and outside of high-MTBE use areas also were compared seasonally (Figure 5.33). Eleven of the river/stream intakes were inside high-MTBE use areas and 6 were outside of these areas. Similar to reservoir/lake source waters, MTBE was detected more frequently and concentrations were greater in high-MTBE use areas than in low-MTBE use areas. However, because as few as 3 detections of MTBE were found in river/stream sources in low-MTBE areas, statistical comparisons were not completed.

Gasoline oxygenates were analyzed in relation to watercraft use as reported in an ancillary data form completed by personnel at each CWS. The average number of watercraft per reservoir per year was divided by the surface area (in acres) of each respective reservoir to obtain a watercraft use factor (number of boats per acre per year). In order to characterize the effect of watercraft on the occurrence and concentration of gasoline oxygenates, the sum of the average detected concentration of all gasoline oxygenates for each reservoir was plotted relative to the watercraft use factor in high- and low-MTBE use areas (Figure 5.34).

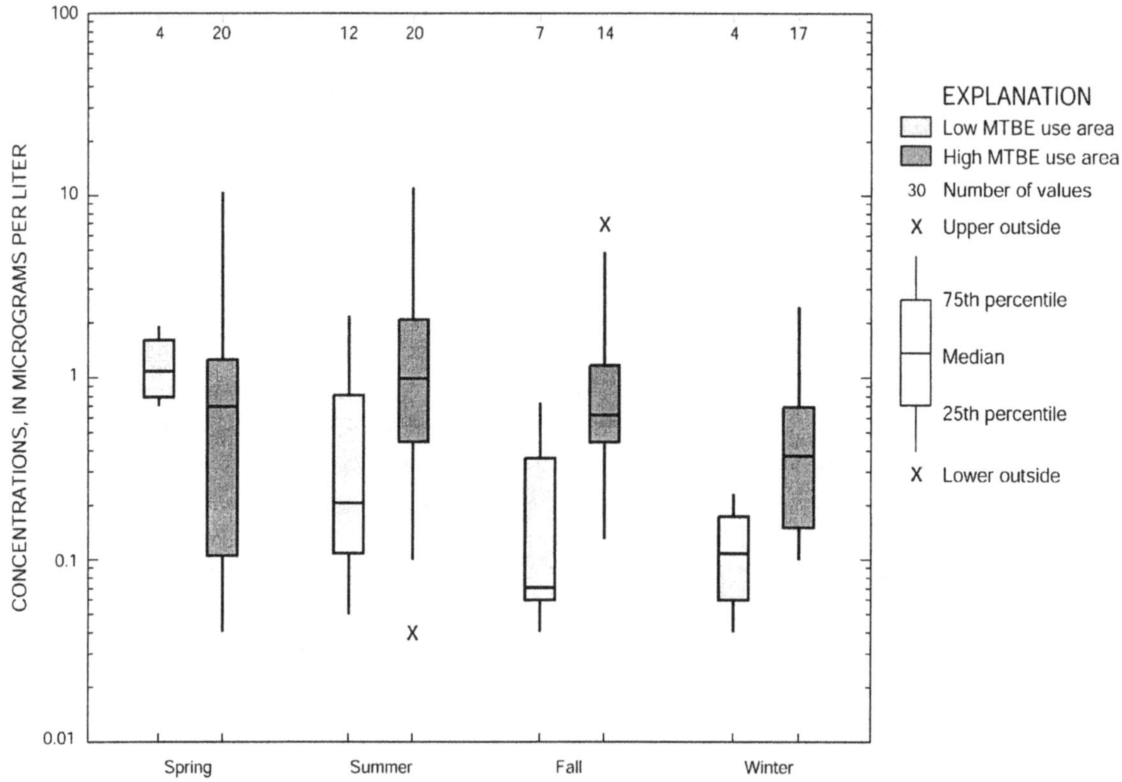

Figure 5.32 Detected concentrations of MTBE in reservoir and lake source waters by location in low- and high-MTBE use areas – Focused Survey

Linear regression analyses indicated a weak positive trend associated with the average gasoline oxygenate concentration and watercraft use in high-MTBE use areas. The relation between the average gasoline oxygenate concentration and watercraft use in reservoirs inside high-MTBE use areas resulted in an r^2 of 0.3783. However, no relation was observed for reservoirs in low-MTBE use areas (r^2=0.0242). In general, as watercraft use increases in areas of high-MTBE use, the average concentration of gasoline oxygenates also increases.

All watercraft use values (inside and outside of high-MTBE use areas) were then divided into four quartile categories: (1) 0 to 25th percentile; (2) 26th to 50th percentile; (3) 51st to 75th percentile; and (4) 76th to 100th percentile. The average gasoline oxygenate concentrations were then plotted per quartile category (Figure 5.35). Again, the average concentration of gasoline oxygenates increased as watercraft use increased.

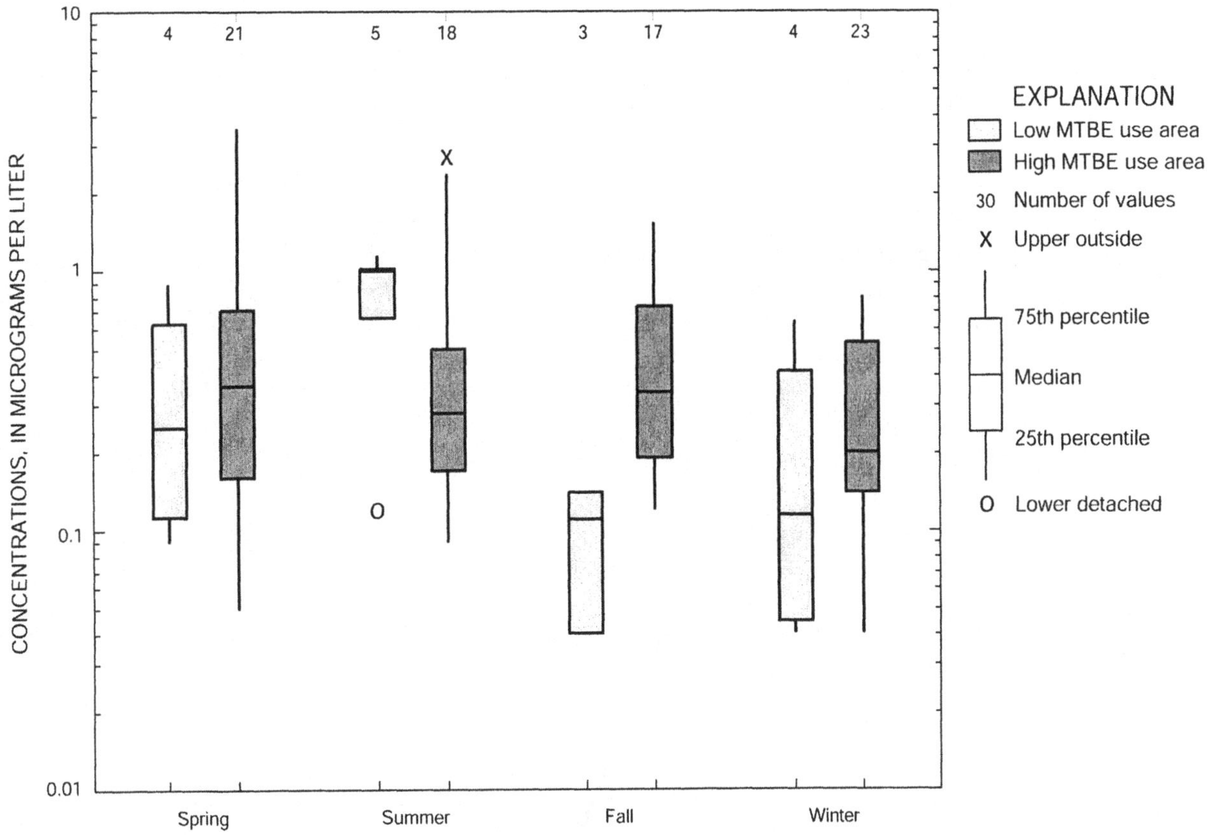

Figure 5.33 Detected concentrations of MTBE in river and stream source water by intake location in low- and high-MTBE use areas - Focused Survey

Trihalomethanes

Trihalomethanes were detected in 11 of the reservoirs/lakes sampled in the Focused Survey. The most frequently detected THM was chloroform, which was found in 10 of the 39 reservoirs/lakes (25.6 percent), followed by bromodichloromethane, which was found in 7 of the 39 reservoirs/lakes (17.9 percent). Seven reservoirs/lakes serving CWSs contained detectable concentrations of at least two THMs. THMs commonly are referred to as disinfection by-products because they can occur in drinking water that has been treated with chlorine. Because samples in the Focused Survey were collected directly from the source water—not finished drinking water—each of these seven CWSs were contacted to better understand the occurrence of these compounds and each provided input to the occurrence and likely source of THMs in their source water.

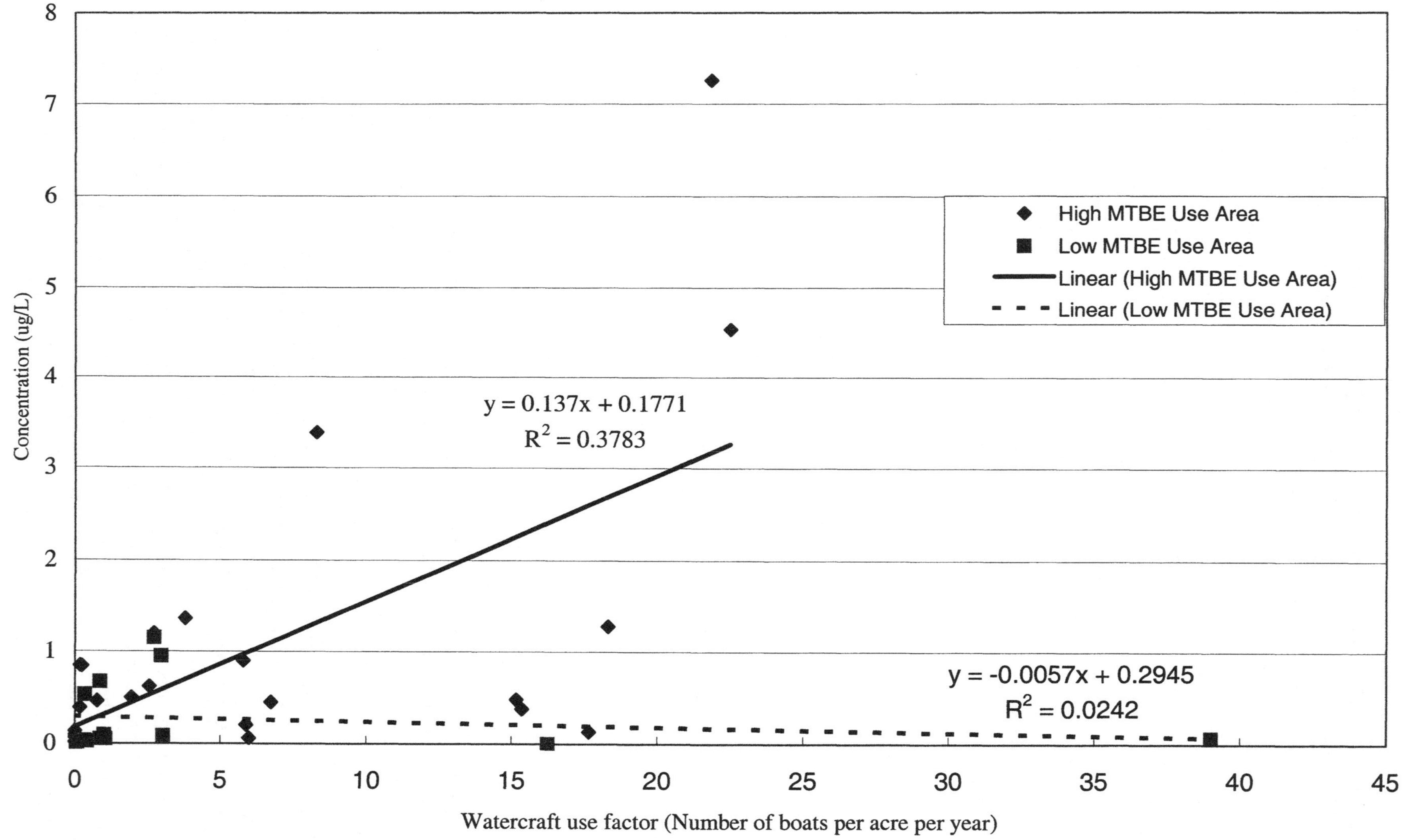

Figure 5.34 Average concentration of sum of gasoline oxygenates versus watercraft usage - Focused Survey

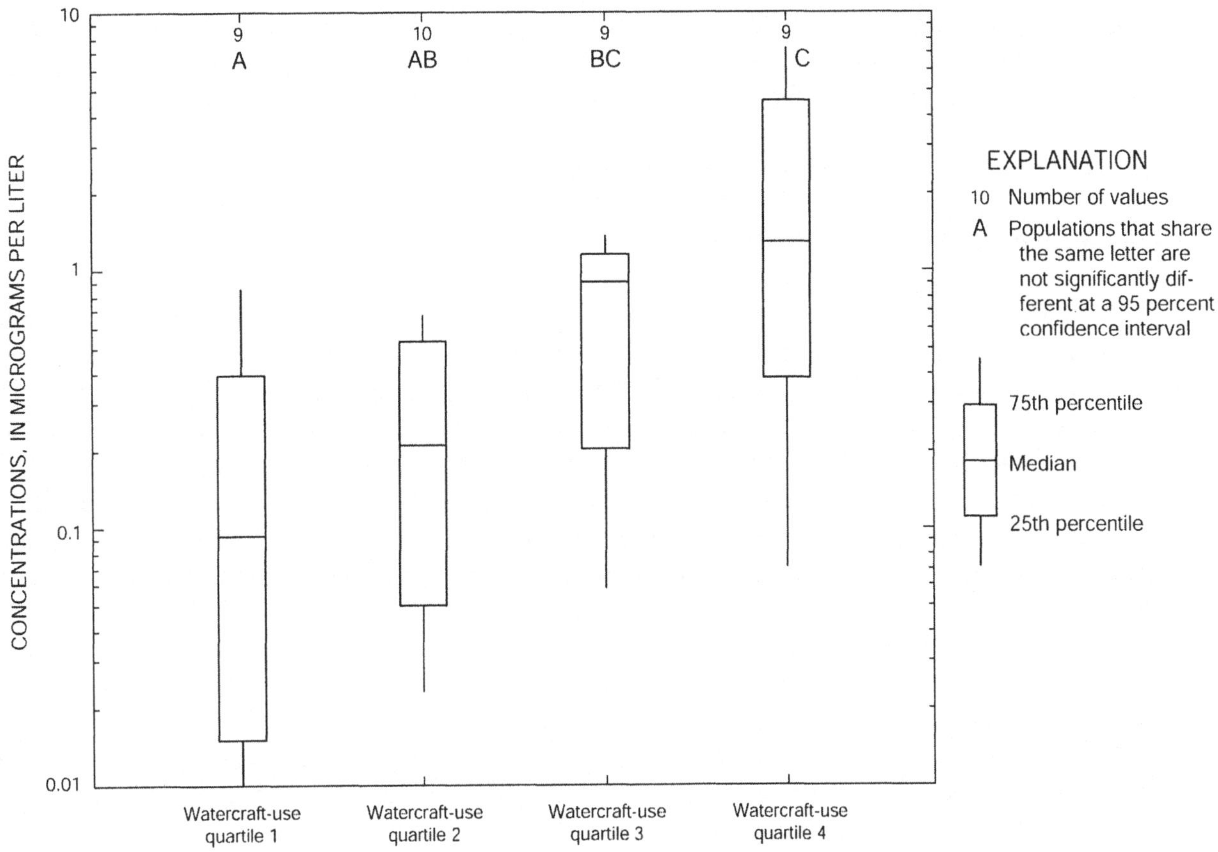

Figure 5.35 Average concentrations of gasoline oxygenates by watercraft-use quartiles – Focused Survey

CWS personnel indicated that THMs were present in their source water as disinfection by-products. Three of the CWSs return filter backwash to the source-water reservoir; one disinfects water from other sources with chlorine and then releases the disinfected water to the reservoir; one CWS adds chlorine upstream of the source reservoir for algae control; one CWS has a powerplant that withdraws water from the reservoir, disinfects the water with chlorine for algal control, uses the water for cooling purposes, and then releases the water back to the reservoir. Lastly, one CWS indicated that the source of THMs in the source water could be due to a wastewater treatment facility discharging upstream of the reservoir.

THMs were detected in 11 of the 17 rivers/streams sampled in the Focused Survey. Chloroform was the most frequently detected THM, occurring in 11 of the 17 rivers/streams (64.7 percent). Nine of these 11 rivers/streams contained at least two THMs.

The CWSs that obtain their source water from the rivers/streams with more than one detectable concentration of any THM were not contacted for additional information to better understand the occurrence of the THMs. It is presumed, however, that the sources of these compounds may include point-source discharges upstream from the CWS intakes. Discharges could include those from wastewater treatment plants, power plants, and industrial discharge. Non-point sources, such as runoff of treated irrigation water from lawns, inflow of ground water containing THMs, and combined sewer overflows that are chlorinated prior to discharge, also could explain the occurrence of these compounds.

FOCUSED SURVEY – MTBE DEGRADATION BY-PRODUCTS

Occurrence of *tert*-Butyl Alcohol (TBA)

The Focused Survey portion of this project provides the first substantial data set on TBA occurrence in environmental water. TBA was detected in 138 of the 273 samples collected (50 percent), which is similar to the detection frequency for MTBE (55.5 percent) observed in the Focused Survey (Table 5.12). The primary TBA results are summarized in Table 5.14 and 5.15, and in Figure 5.36 (A-D) versus four category variables: sampling season (A), source type (B), source size (C), and source state (D). Almost all categories show median values of TBA concentration that are slightly greater than the method MRL of 0.23 μg/L. There are no strong differences in the data by category, but several weak patterns are suggested by the data:

1. Very high TBA concentrations in a few samples as well as the highest value were observed in the spring (Figure 5.36 A). However, the strong seasonal occurrence that is sometimes observed with MTBE is not evident for TBA.
2. Very high TBA concentrations in a few samples as well as the highest median value were observed in well water (Figure 5.36 B). The various types of surface water supplies showed uniformly low values of TBA.
3. The highest median TBA concentrations were observed in the small and large CWSs (Figure 5.36 C).
4. The statistical distribution of TBA occurrence data varied widely with state (Figure 5.36 D), with the highest median and/or extreme values from Georgia, Illinois, Kansas, and New Jersey. Maryland had one very high value but a low median value. To check for regional trends in TBA concentrations, the data were plotted on a map of the contiguous states (Figure 5.37), but it reveals no strong patterns.

Occurrence of Other Possible Oxygenate Degradation Products

Tertiary-butyl formate (TBF) was not detected in any environmental sample or blank. However, TBF is not stable in the heated purge and trap gas chromatography/mass spectrometry (HP&T-GC/MS) method, and is partially converted to TBA. From blanks spiked with TBF, it is known that TBF would be detected by the method if the sample contains >0.25 μg/L, therefore, it can be concluded that TBF levels were below 0.25 μg/L in all samples analyzed in the Focused Survey. Similarly, it is not known whether some of the TBA detected by HP&T/GC-MS was formed by hydrolysis of TBF during analysis, but it is known that this effect could not have increased the measured TBA concentrations by more than 0.25 μg/L, which would have had little effect on the results presented below. Furthermore, since any TBF is likely to arise as an intermediate in degradation of MTBE to TBA (Figures 1.2 and 1.3 – Chapter 1), this ambiguity is of little consequence in the discussion that follows.

Tert amyl alcohol (TAA) was not a focus of the study, but was noted a few times during the peak identification process (checking mass spectra) because it elutes very near TBF. In few cases were the peaks large enough to be good candidates for quantification. Other possible metabolites of MTBE (or other oxygenates) include 2-methyl-2-hydroxy propanol, isopropanol, and acetone. Although HP&T/GC-MS might be a good method for detecting low levels of these compounds, the data analysis was not optimized for these compounds during this study because they (*i*) are likely to be of little environmental consequence, (*ii*) are too biodegradable to make good indicators of MTBE degradation, and (*iii*) would have detracted from the number of TBA analyses that could be performed.

Table 5.14
Summary of results for *tert*-butyl alcohol (TBA) – Focused Survey

	No. of Samples Analyzed*	No. of Detections†	Detection Frequency (%)	Median Value (µg/L)	Highest Value (µg/L)
Season					
Winter	49	28	57	0.28	7.9
Spring	92	47	51	0.34	200.9
Summer	93	38	41	0.28	2.35
Fall	39	25	64	0.37	1.30
Type					
Well	93	32	34	0.65	201
Spring	1	0	0	-	-
Irrigation	5	3	60	0.43	1.45
Canal	6	4	67	0.34	0.48
Stream	10	8	80	0.36	0.88
River	63	41	65	0.29	1.32
Lake	11	3	27	0.21	0.29
Reservoir	84	47	56	0.27	2.35
Size					
Very small	12	1	8	0.41	0.41
Small	25	10	40	3.72	51.6
Medium	22	9	41	0.26	3.72
Large	35	15	43	0.42	200.9
Very large	179	103	57	0.29	46.8

* Environmental samples only, including duplicates.
† Those not suspected of contamination and above the MDL.

Table 5.15
Summary of results for *tert*-butyl alcohol (TBA) by State – Focused Survey

State	No. of Samples Analyzed*	No. of Detections†	Detection Frequency (%)	Median Value (μg/L)	Highest Value (μg/L)
AZ	7	3	43	0.43	1.45
CA	49	31	63	0.27	1.52
CO	3	0	0	-	-
CT	7	2	29	0.20	0.26
DE	22	11	50	0.49	0.88
FL	9	7	78	0.16	0.22
GA	6	5	83	2.33	201
IA	3	3	100	0.25	1.19
IL	4	3	75	6.26	7.94
IN	1	0	0	-	-
KS	6	4	67	2.19	51.6
KY	8	4	50	0.37	0.61
LA	2	2	100	0.42	0.64
MA	8	1	12	0.37	0.37
MD	8	7	87	0.39	20.7
ME	5	1	20	0.25	0.25
MN	6	1	17	0.36	0.36
MS	3	2	67	0.25	0.25
NC	4	2	50	0.19	0.29
NH	6	2	33	0.20	0.21
NJ	10	5	50	1.37	46.8
NV	3	0	0	-	-
NY	8	3	37	0.16	5.57
OH	1	0	0	-	-
OK	6	5	83	0.48	1.08
OR	13	3	23	0.13	0.14
PA	18	13	72	0.49	1.32
RI	2	1	50	0.09	0.09
SD	2	1	50	0.29	0.29
TN	5	3	60	0.29	0.42
TX	22	6	27	0.44	1.30
VA	12	7	58	0.23	9.28
VT	2	0	0	-	-
WV	2	0	0	-	-

* Environmental samples only, including duplicates.
† Those not suspected of contamination and above the MDL.

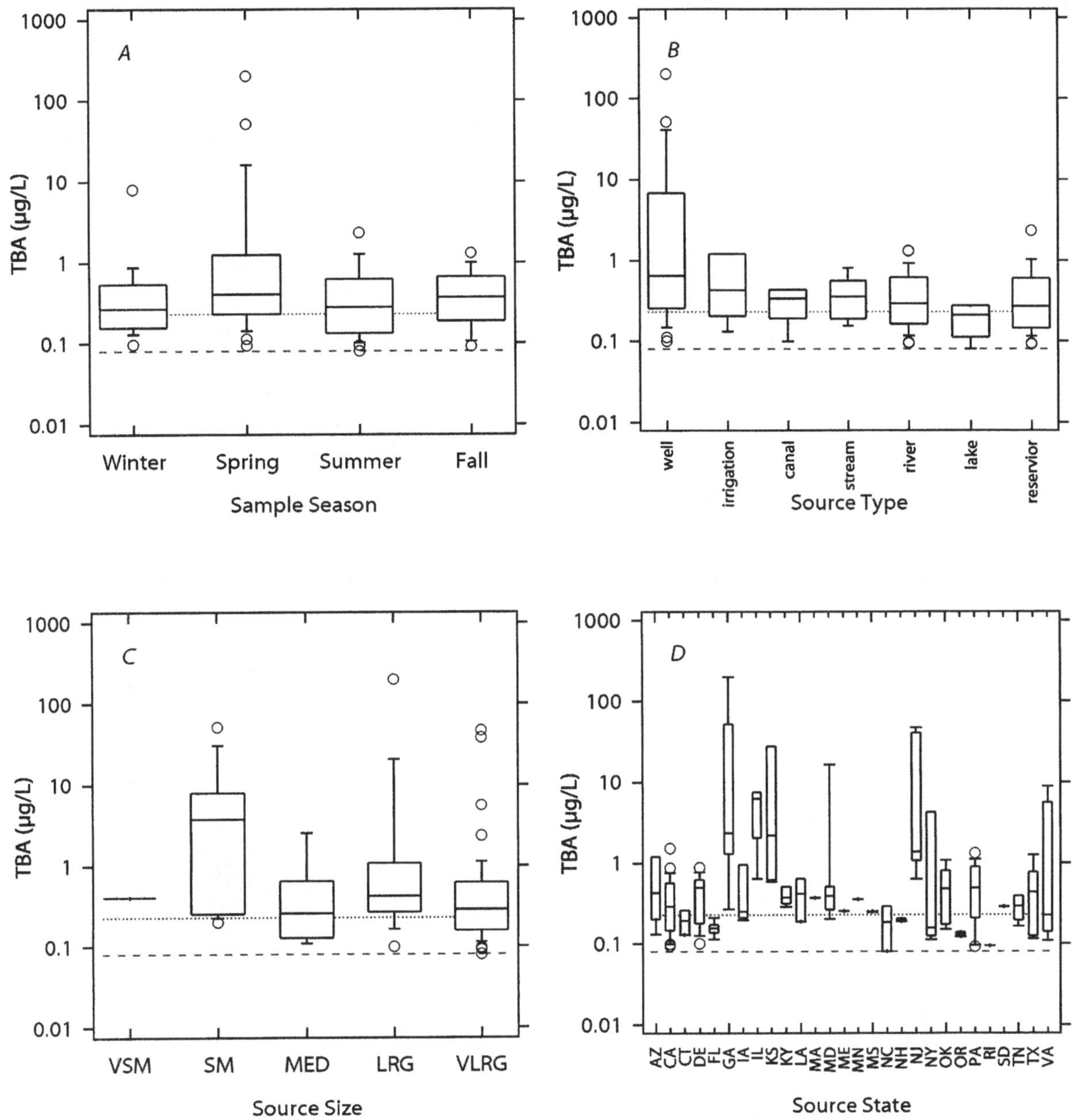

Only environmental sample data with TBA above the MDL and without suspected contamination are shown.

MDL- minimum detection level; MRL- minimum reporting level

Box – represents 25th to 75th percentile. Line in box is median.

Whiskers – represent 10th to 90th percentile.

Circles represent points outside the 5th and 95th percentile.

Figure 5.36 Results for TBA, box-and-whisker plots representing the distribution of measured concentrations vs. several category variables - Focused Survey

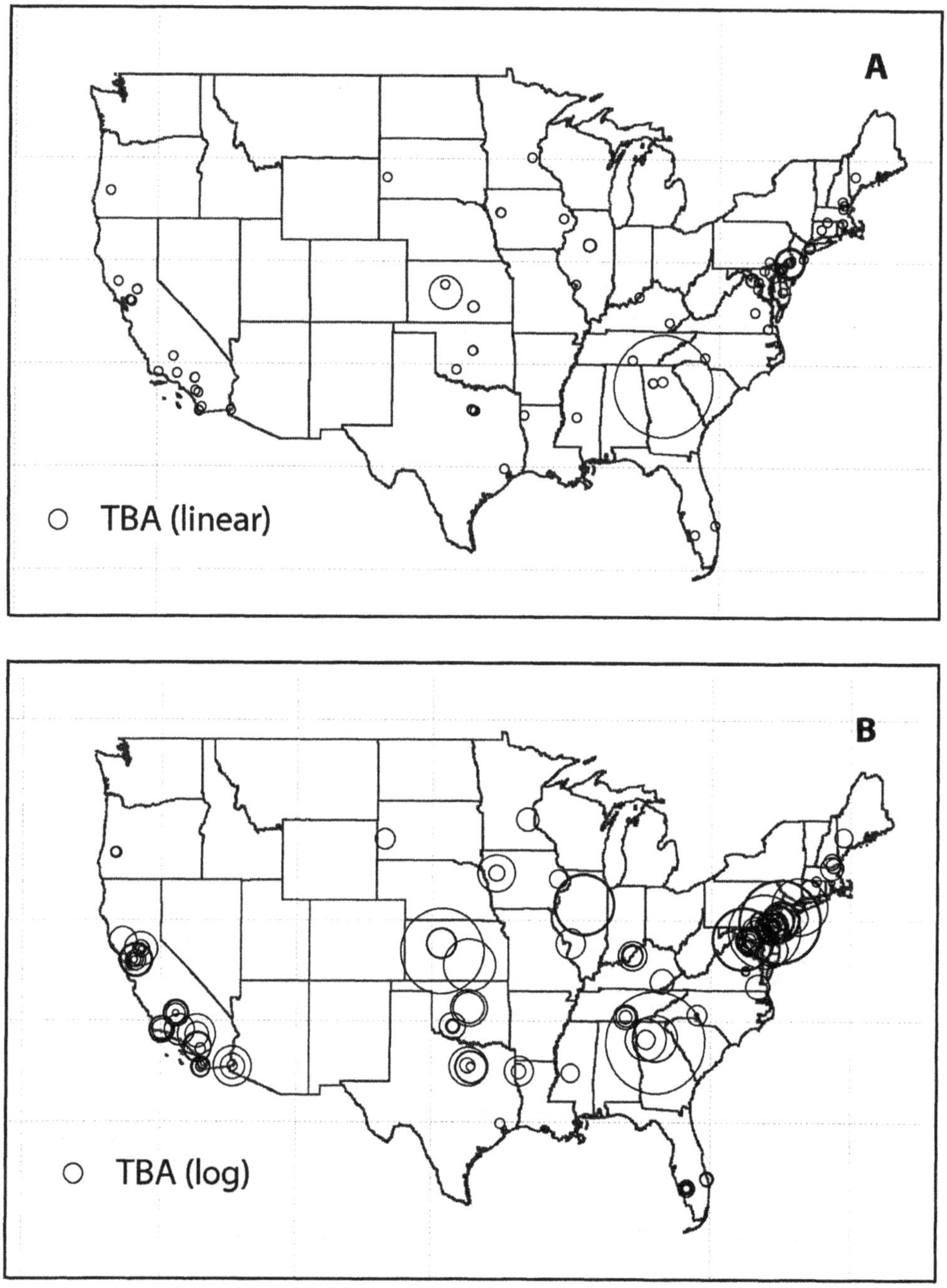

The TBA concentration is represented by the size of the circles: A-scaled linearly, B-scaled logarithmically.

Only environmental sample data with TBA above the MDL and censored for random contamination are shown

Figure 5.37 Results for TBA, showing the geographic distribution of the data - Focused Survey

Co-occurrence of TBA, MTBE, and BTEX

To provide a comprehensive perspective of the co-occurrence of TBA, MTBE, and the major BTEX constituents (benzene, toluene, ethylbenzene, and xylenes), the combinations of these compounds are presented as a matrix of scatter plots (Figure 5.38). As expected, the constituents of BTEX generally show high correlations. TBA and MTBE also show a significant degree of correlation, as will be discussed further below. It is notable, however, that the data for TBA and MTBE do not show any strong correlations with the data for BTEX compounds. Note that all of the data in Figure 5.38 are scaled logarithmically, which gives more uniform distributions of the data and therefore less leverage to the few very large values relative to the many values that are just above the MRL.

The comparison between TBA and MTBE detections is of particular interest in this portion of the study. Figure 5.39 shows all TBA data versus all the data for MTBE obtained by OHSU on the same samples (i.e., both data sets prior to any censoring). Where an environmental sample was provided to OHSU in duplicate, the resulting measured values of TBA and MTBE are connected with a line. In most cases, the duplicates plot very closely, and the connecting line is barely evident. This presentation highlights the few cases of duplicate pairs with notable different values of TBA and/or MTBE. Most of the lines are more vertical than horizontal, indicating that the notable differences among replicates arose mostly for TBA. The few cases where the bars represent 5-fold or greater differences all involve TBA or MTBE values near or below the MRL.

Censoring the TBA and MTBE data for possible random contamination and at the MDL is shown in Figure 5.40. Qualitatively, there appears to be a scattered but positive correlation between measured values of TBA and MTBE (i.e., samples with high MTBE also tend to have high TBA). The degree of correlation between these two variables is measured by the linear correlation coefficient (Pearson's r), and $r = 0.59$ for the (log transformed) data in Figure 5.40.

For $n = 95$, an $r = 0.59$ is a significant correlation at the 99.95% confidence level. Least-squares, linear regression on the data gives $\log[TBA] = (0.49 \pm 0.27) \log[MTBE] - (0.37 \pm 0.22)$ where the uncertainties represent 95% confidence intervals in the parameter estimates. Thus, at this confidence level, the slope of the correlation between TBA and MTBE concentrations is greater than 0 but less than 1, and the intercept is less than 0. The significance of these parameters will be discussed further in the next section.

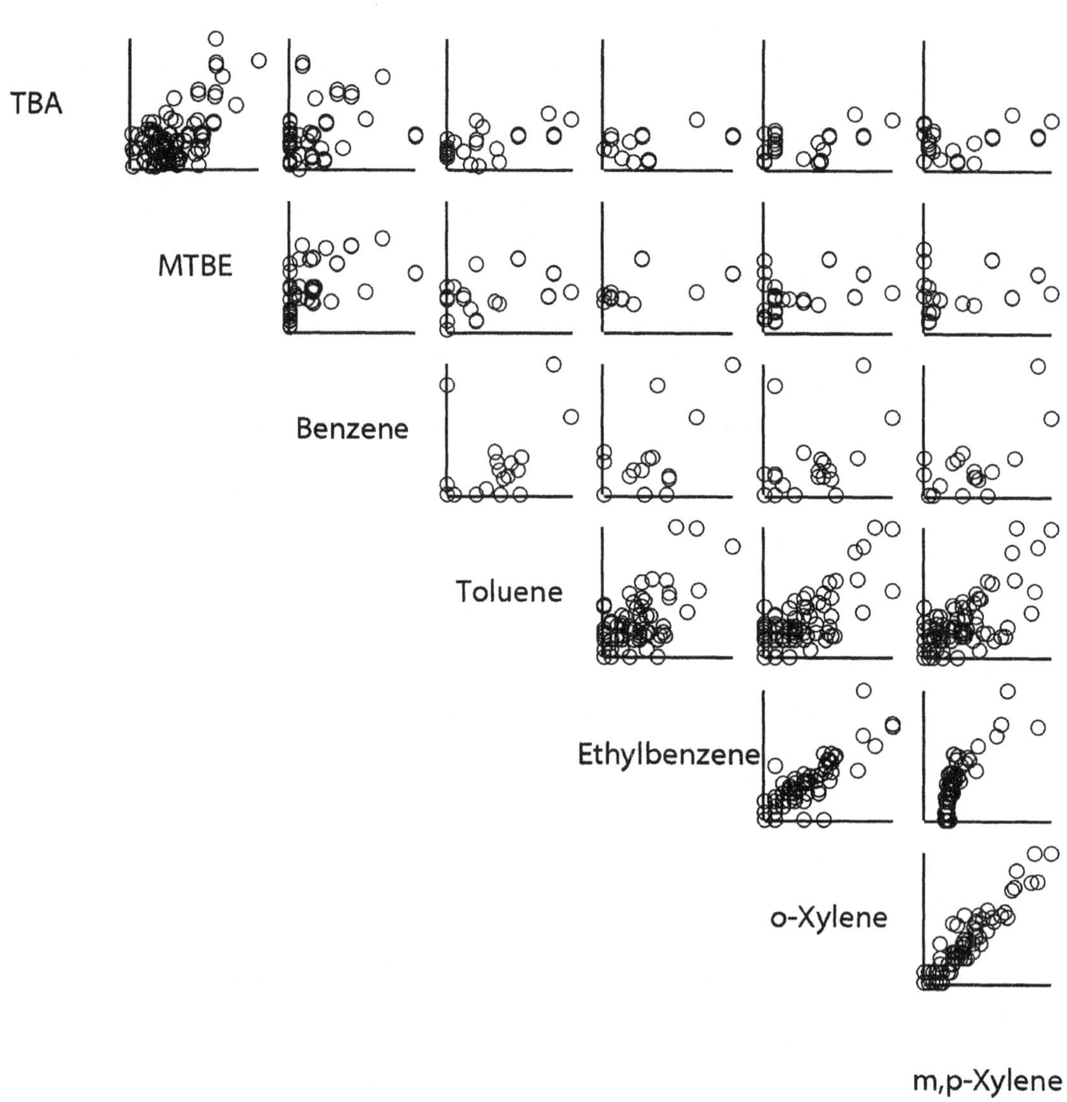

All axes represent log concentration, as seen in other figures.
The axis range is constant across each row and column.

Figure 5.38 Matrix of scatter plots for all combinations of TBA (by OHSU), MTBE (by OHSU), and the constituents of BTEX (by MWDSC) – Focused Survey

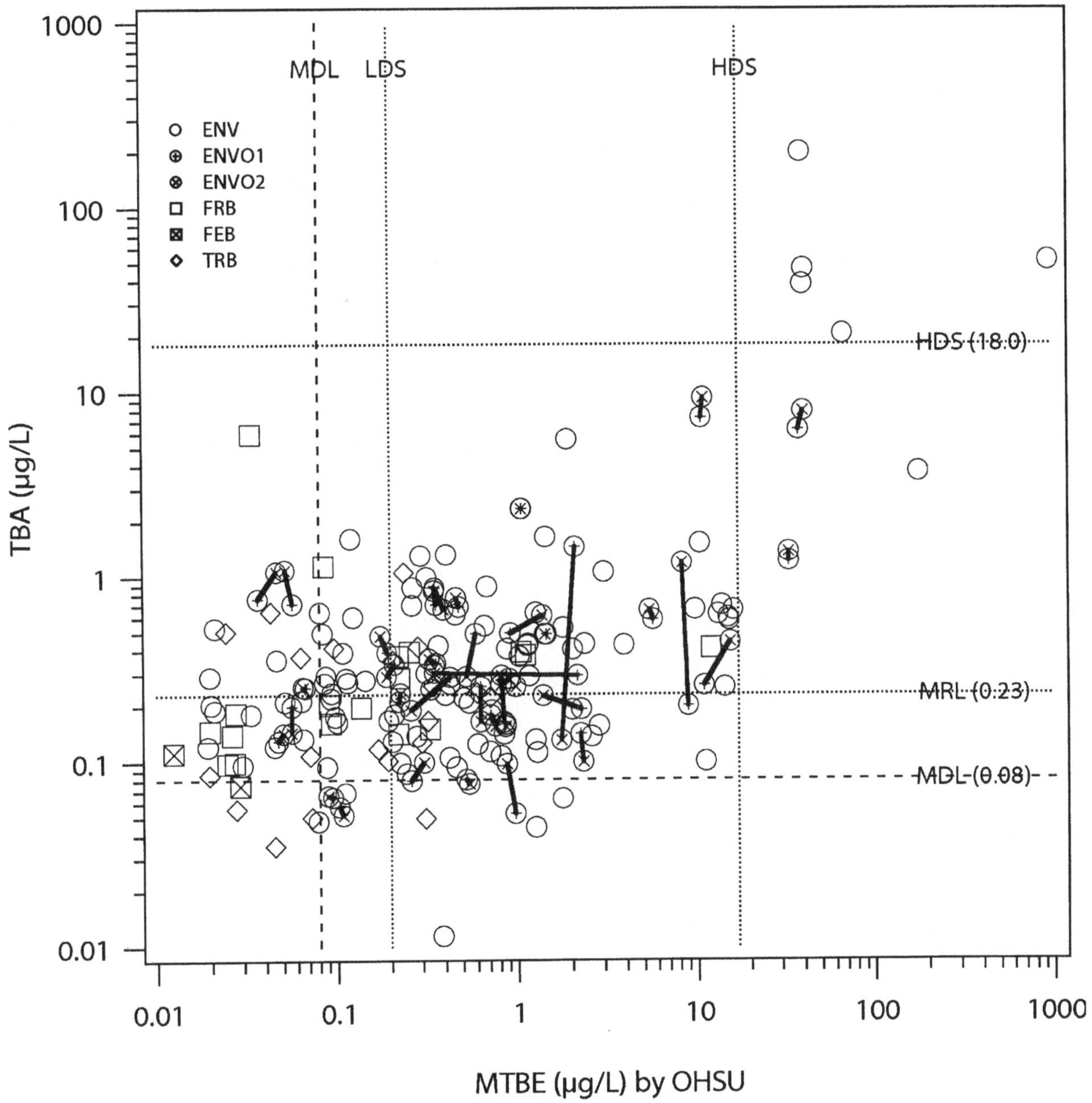

All detected values are shown (including duplicates, and blanks).
MDL - the minimum detection level
MRL - minimum reporting level
HDS - highest daily standard
LDS – low daily standard
MTBE data are those from OHSU

Figure 5.39 Results for TBA vs. MTBE, with duplicates connected to show agreement - Focused Survey.

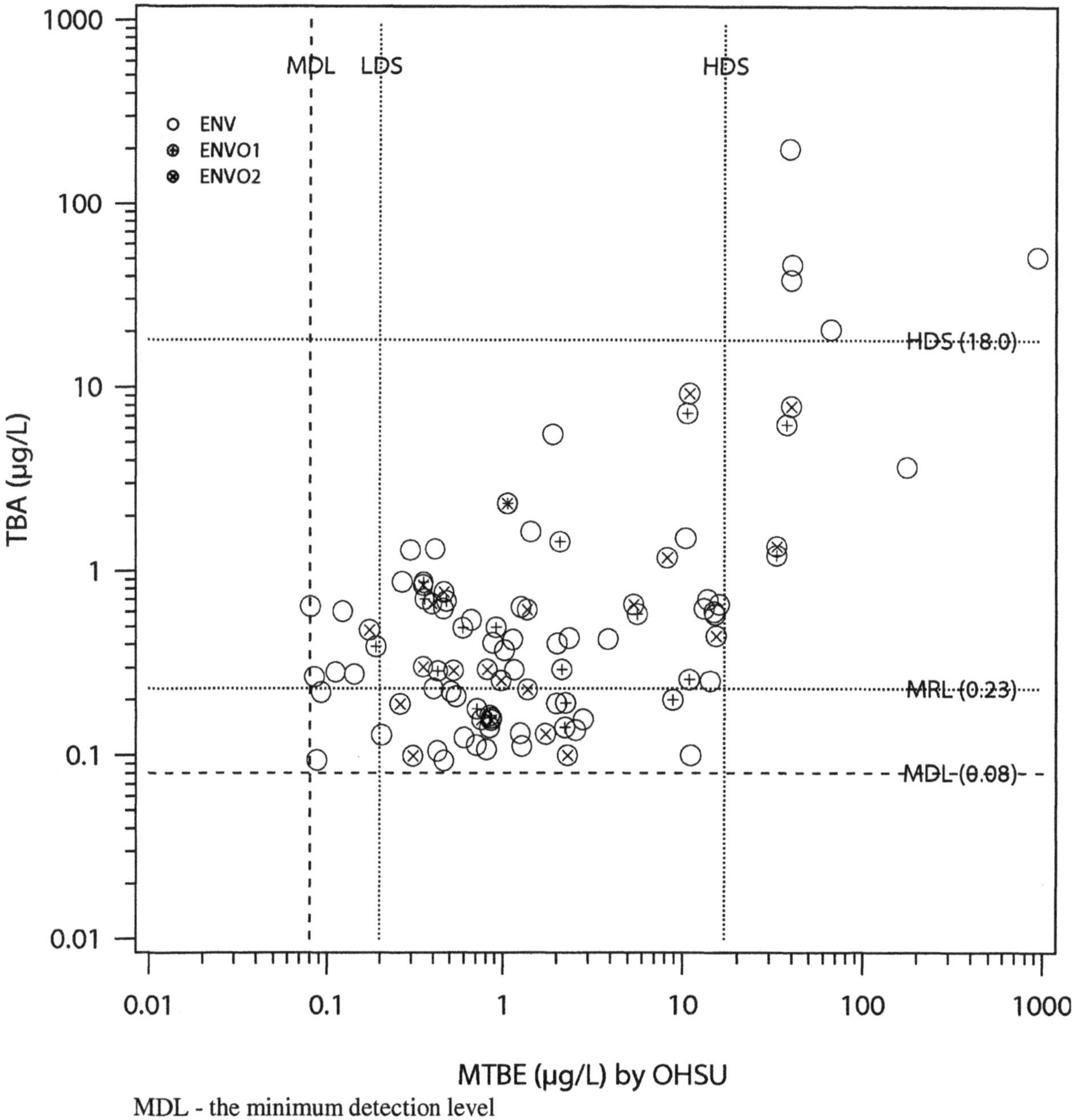

MDL - the minimum detection level
MRL - minimum reporting level
HDS - highest daily standard
LDS – low daily standard
MTBE data are those from OHSU

Figure 5.40 Results for TBA vs. MTBE, showing only environmental sample data with TBA and MTBE above the MDL and censored for random contamination - Focused Survey

One way to derive additional information from Figure 5.40 is to apply a third, potentially-explanatory variable as shown in Figure 5.41 and Figure 5.42. The former shows that the samples with high TBA and MTBE tend to be well waters, and that most samples that were contaminated with TBA and MTBE only near their MDLs were from surface waters. Figure 5.42 is equivalent to Figure 5.41 but with the symbol size reflecting the size of the utility from which the sample was collected. It shows that small water supplies make up most of the data with high TBA and MTBE concentrations.

Co-occurrence (TBA/MTBE) vs. Other Explanatory Variables

To further investigate the relationship between concentrations of TBA and those of MTBE, we have calculated the concentration ratio (TBA/MTBE) and plotted it versus a variety of potential-explanatory variables. Figure 5.43 shows TBA/MTBE versus two temporal variables: the sample collection date and the sample storage time (date sampled minus date sample was analyzed). Systematic changes in TBA/MTBE with sample collection date (Figure 5.43 A) might reveal a gradual environmental transformation of MTBE and/or TBA. While systematic changes in TBA/MTBE with storage time (Figure 5.43 B) might suggest transformation of the contaminants during storage (and therefore potential for degradation in the field). However, the data appear to be uniformly distributed with respect to these temporal variables, and therefore do not provide evidence for any transformation of MTBE to TBA within the timeframe of this study. Furthermore, coding the data by the sample source type (different symbols in Figure 5.43) does not reveal any strong trends within subsets of the data.

In Figure 5.44, TBA/MTBE is represented versus the same four category variables used in Figure 5.36 (i.e., sampling season, source type, source size, and source state). The slope of the regression line for the data shown in Figure 5.40 is included in these box plots (horizontal dashed lines) to represent the global mean value of TBA/MTBE.

Another useful basis for comparison is the value of TBA/MTBE that is representative of gasoline formulations used in the areas that were sampled in this portion of the study. The data are not available to determine this value with confidence, so it is not shown in Figure 5.44. However, from data compiled by the National Institute for Petroleum and Energy Research (NIPER), it appears TBA/MTBE is <0.01 for the vast majority of gasoline samples where both compounds were detected. Thus, even allowing for considerable uncertainty and variability in TBA/MTBE for gasoline, this ratio would be considerably less than most values of TBA/MTBE obtained from samples in this study.

.

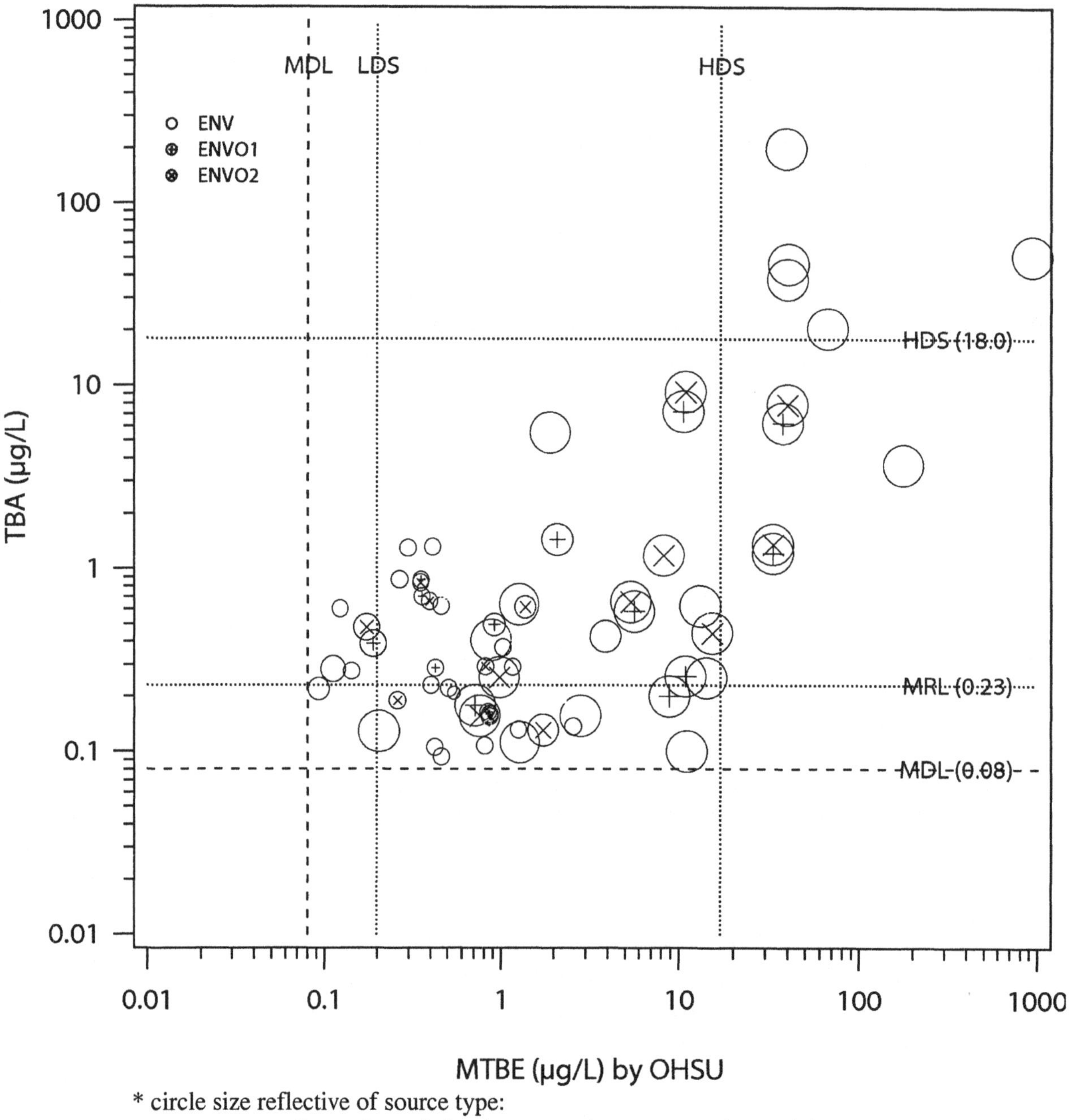

* circle size reflective of source type:

well > spring > irrigation > canal > stream > river > lake > reservoir -

Only environmental sample data with TBA and MTBE above the MDL and censored for random contamination are shown.

MTBE data are those from OHSU

Figure 5.41 Results for TBA vs. MTBE, with symbol size coded to source type – Focused Survey

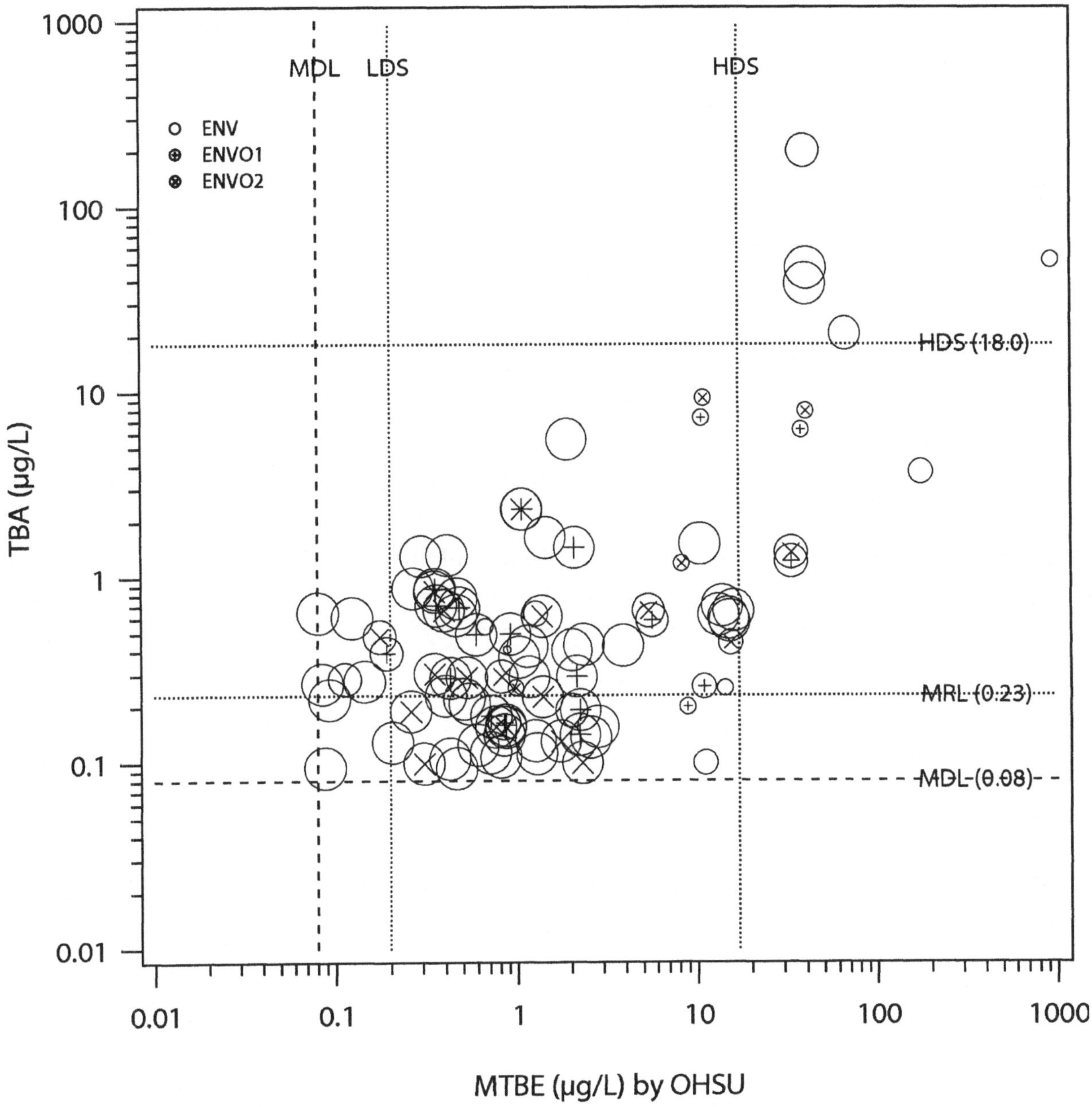

* Large points correspond to large supplies and small points correspond to small supplies.

Only environmental sample data with TBA and MTBE above the MDL and censored for random contamination are shown.

MTBE data are those from OHSU

Figure 5.42 Results for TBA vs. MTBE, with symbol size coded to source size* – Focused Survey

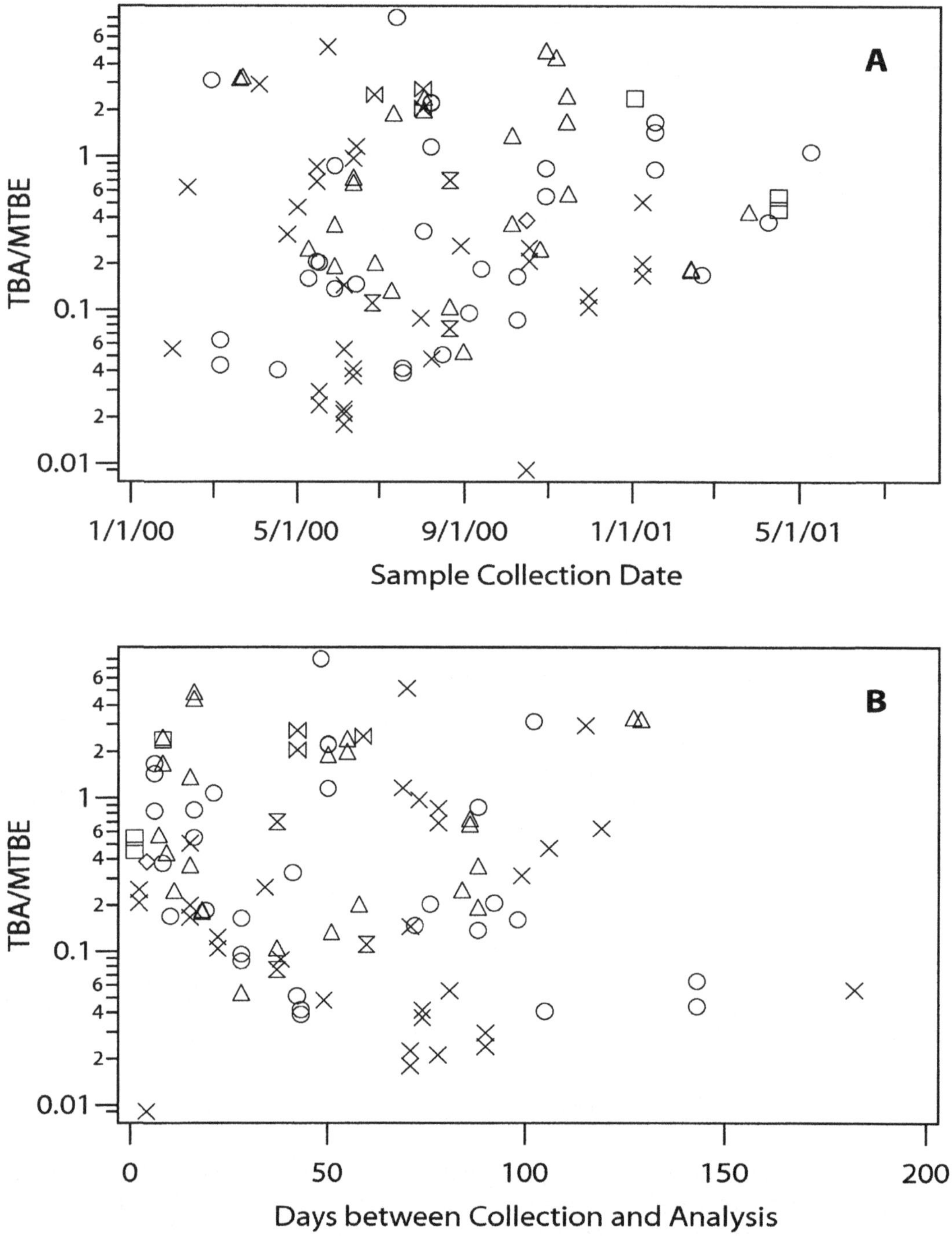

Only censored data are shown.
X– Well, -* Spring, , □- stream, Δ- river, ◊- lake, and ○- reservoir

⧖ - Irrigation

Figure 5.43 Ratio of TBA to MTBE vs. (A) sample collection date and (B) days elapsed between (nominal) sample collection data and the date of sample analysis – Focused Survey

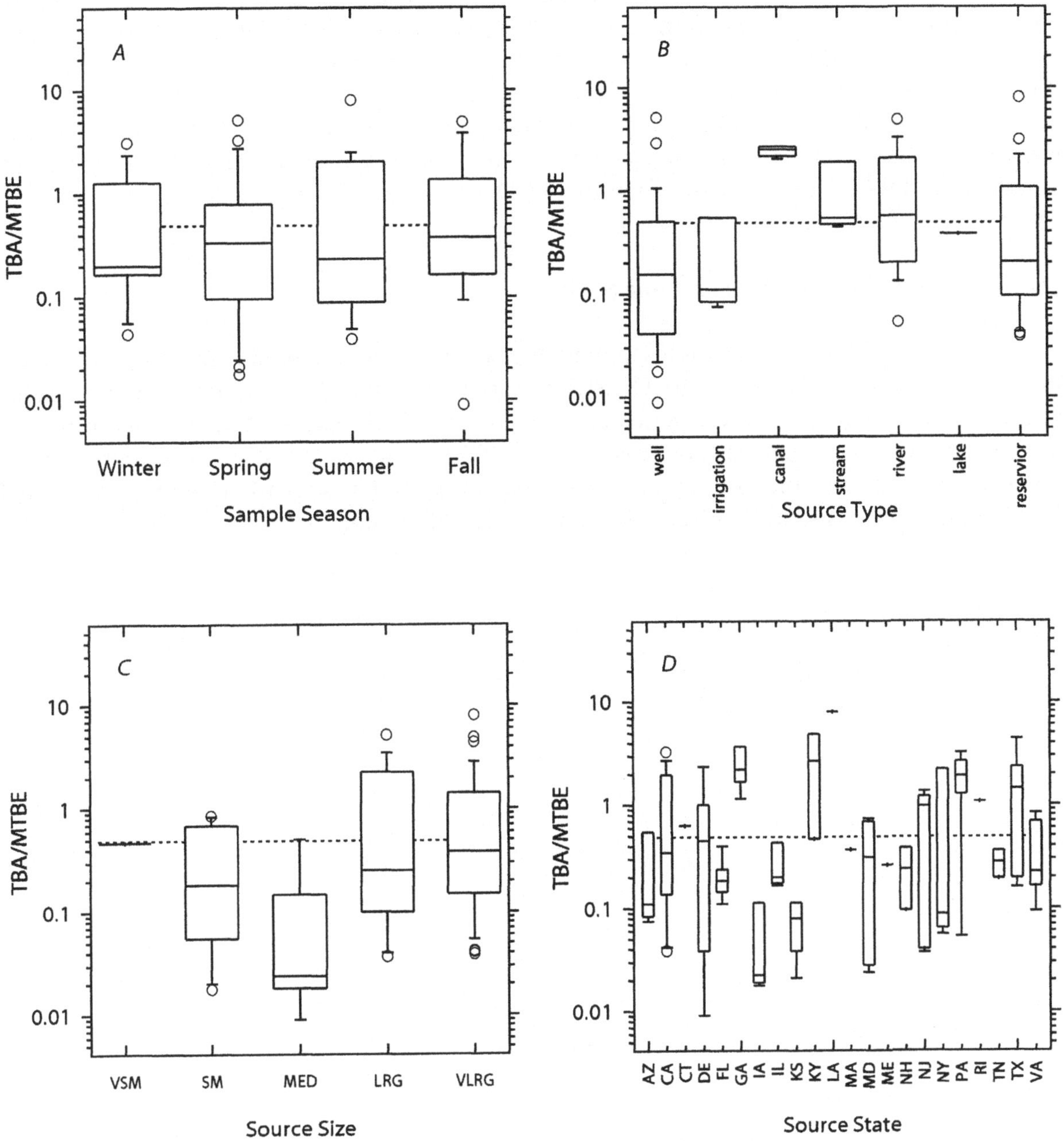

Only environmental sample data with TBA and MTBE above the MDL and censored for random contamination are shown.

Box – represents 25th to 75th percentile. Line in box is median.

Whiskers – represent 10th to 90th percentile.

Circles represent points outside the 5th and 95th percentile

Figure 5.44 Results for TBA/MTBE, box-and-whisker plots representing the distribution of measured concentrations vs. several category variables - Focused Survey

Since all of the data in Figure 5.44 fall well above the value of TBA/MTBE that might be expected in gasoline (based on the NIPER survey data), it is clear that the observed environmental concentrations of TBA are enriched relative to MTBE. This might arise due to (*i*) degradation of MTBE to TBA, (*ii*) differential dissolution and transport of TBA relative to MTBE, or (*iii*) additional sources of environmental TBA other than gasoline.

The data in Figure 5.44 might provide preliminary evidence for one or another of these possibilities, but the four types of categories shown do not produce any strong, systematic variations in TBA/MTBE. Finally, Figure 5.44 D and Figure 5.45 provide no patterns in the distribution of TBA/MTBE by state or region.

Co-occurrence (TBA/MTBE) as Time Series

There were 20 sites that were sampled three or more times over the course of the Focused Survey, and samples were sent to OHSU in part to provide data for time series analysis. However, only about 10 of these series produced both MTBE and TBA detections in three or more samples. There were several cases where TBA went up steadily over four sampling periods, but most cases did not show any clear pattern (data not shown). The TBA/MTBE data were remarkably consistent for each CWS (i.e., no sign of temporal trends).

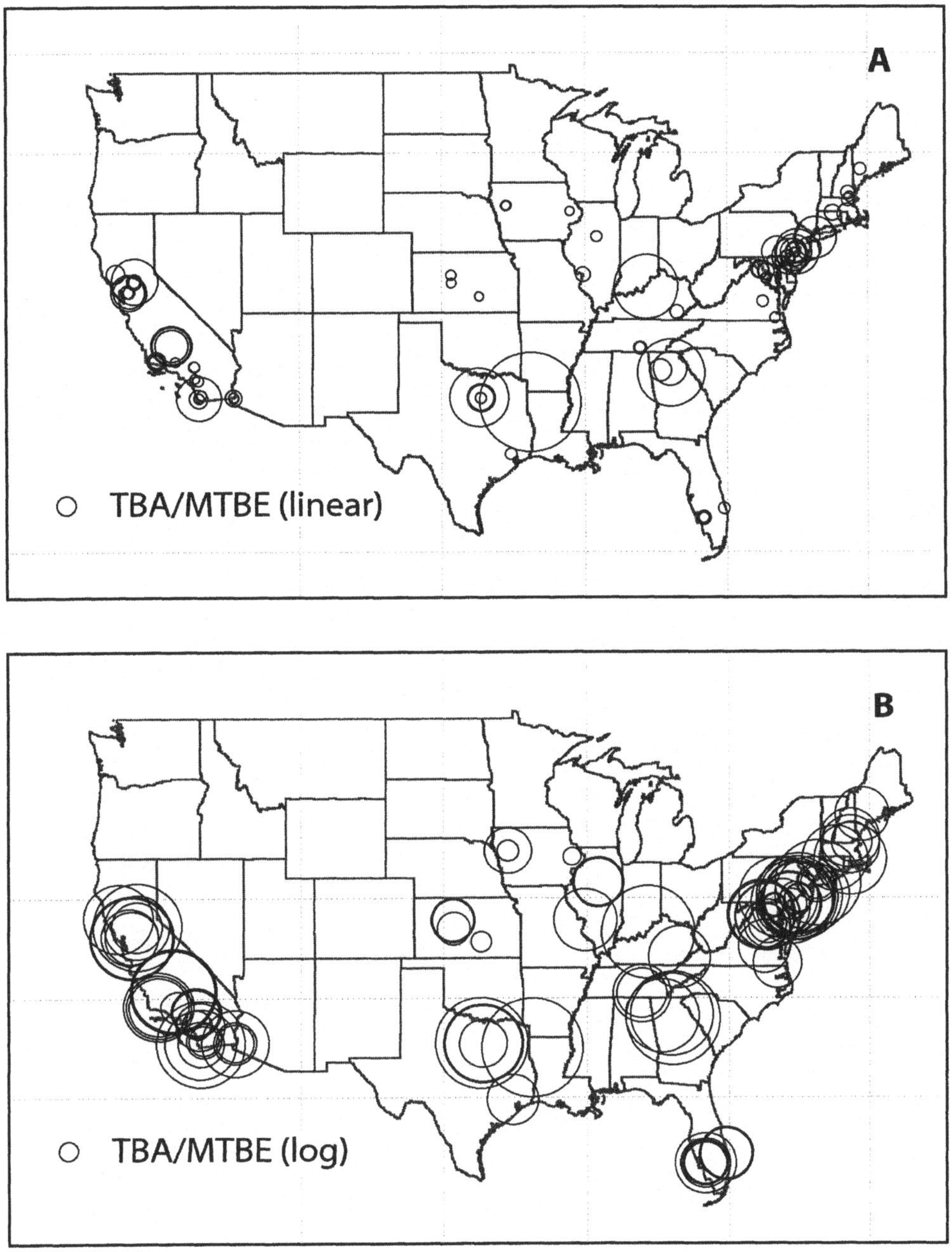

The TBA concentration is represented by the size of the circles: A-scaled linearly, B-scaled logarithmically.

Only environmental sample data with TBA and MTBE above the MDL and censored for random contamination are shown

Figure 5.45 Results for TBA/MTBE, showing the geographic distribution of the data - Focused Survey

CHAPTER 6
SUMMARY AND CONCLUSION

LITERATURE SURVEY

Inconsistencies among assessments reviewed include different objectives, reporting levels, and different water types sampled such as ambient ground water, and treated and untreated drinking water. This usually made comparisons among assessments difficult, if not impossible, to achieve. Regardless, results of this literature review indicate that MTBE has been detected in public and (or) private drinking-water supplies in 36 states at concentrations ranging from 0.1 to 17,800 µg/L (micrograms per liter). However, when median detected concentrations were available, they were typically low--less than 5.0 µg/L. In addition, the reviewed assessments collectively indicate that: (1) MTBE occurred in public drinking-water systems supplied by ground and surface water, and concentrations generally were less than 20 µg/L; (2) population density and reformulated gasoline use were significant risk factors for MTBE detection in water supplies; and (3) type of well, water supply, and proximity to gasoline storage tanks did not seem to be associated with MTBE detection.

A review of existing information on the aesthetic effect of MTBE in drinking water was also conducted. The odor has been described as a sweet solvent (such as turpentine). Although a definitive threshold level has not been well defined, the information indicates that low levels, ranging from 14 to 71 µg/L can cause an unpleasant aroma and flavor to the water. This range is congruous with the EPA advisory range of 20 to 40 µg/L. However, the state of California has set a secondary standard of 5 µg/L.

RANDOM SURVEY

The Random Survey was designed to provide representative information on the frequency of detection, concentration, distribution, and temporal variability of MTBE, other ether gasoline oxygenates, and other VOCs in source waters used by CWSs in the United States. Source-water samples for the Random Survey were collected from 954 CWSs in the 50 states, Native American Lands, and Puerto Rico from May 3, 1999 to October 23, 2000. Samples were allocated proportionally to the total number of systems in each of the five population served-size categories, the total number of people served by each of the source-size category, and the type of source waters used. Untreated source-water samples from 579 ground-water sources (wells, springs, and galleries) and 375 surface-water sources (rivers, aqueducts, canals, lakes, and reservoirs) were collected by CWS personnel and sent to the MWDSC laboratory for analysis of MTBE, 3 other ether gasoline oxygenates, and 62 additional VOCs. One source-water sample was collected and analyzed from each participating CWS.

Forty-two of the 66 VOC analytes were detected in at least one sample at concentrations equal to or greater than the MRL of 0.2 µg/L (for all VOCs except methyl ethyl ketone, which has an MRL of 2.0 µg/L). One or more VOCs were detected in 257 (27 percent) of the 954 source-water samples. Chloroform was the most frequently detected VOC in 13 percent of the samples; MTBE was second, detected in 8.7 percent of the samples. Although VOC detections were frequent, concentrations were generally less than 10 µg/L. Ninety-five percent of all 530

VOC detections were below 10 μg/L—and 63 percent were less than 1.0 μg/L. However, eight source-water samples contained one or more VOCs at concentrations that would have exceeded Federal maximum contaminant levels (MCLs) established for drinking water. When VOCs were detected in source-water samples, co-occurrence of several VOCs was fairly common, with multiple detections in nearly half (47 percent) the 257 samples with VOC detections.

Proportionally, more surface-water samples (30 percent) than ground-water samples (25 percent) contained VOCs. Gasoline compounds collectively and MTBE were detected significantly more often in surface-water sources than ground-water sources at the 95-percent confidence level, whereas, the opposite was true of solvents and refrigerants. For most VOC subgroups, detections were more frequent in samples from the largest water systems as their source waters are more likely to be in urban settings. Detection frequencies for any VOC, non-THM VOCs, BTEX compounds, solvents, and refrigerants were significantly greater in source-water samples from areas with 60 percent or more urban land use than in less urbanized areas. Also, drinking-water sources in areas with population density greater than 1,000 people per square mile contained any VOC, non-THM VOCs, gasoline compounds collectively, MTBE, solvents, and refrigerants significantly more often than source waters in less densely populated settings.

The widespread occurrence of THM disinfectant by-products in source waters probably represents the persistence of residual concentrations of disinfectant by-products from chlorinated drinking water and wastewater circulating through the hydrologic cycle. The more frequent detections of MTBE in surface-water sources (14 percent) than ground-water sources (5.4 percent) probably is related to emissions and leaks or spills of gasoline from personal and commercial motorized watercraft operated on surface-water bodies that are used for drinking-water supply. Detections of MTBE in surface-water samples were most frequent in samples collected during summer months, reflecting the seasonal use of watercraft, whereas detections in ground water did not demonstrate any seasonal effects. Concentrations of MTBE in source-water samples were generally less than 5.0 μg/L. Furthermore, only one sample matched the 20 μg/L lower level of the federal recommended drinking-water advisory. Co-occurrence of MTBE and BTEX was limited to a few samples with low concentrations of both gasoline contaminants. Detections of other ether gasoline oxygenates—ethyl *tertiary* butyl ether, *tert*-amyl methyl ether, and diisopropyl ether—were rare and only occurred with MTBE.

The frequency of detecting MTBE in source-water samples was significantly greater in source waters located in areas of the nation where MTBE is used in high volume in gasoline (23 percent) compared to drinking-water sources outside of these areas (4.4 percent). This five- fold increase in MTBE detections in high MTBE-use areas is consistent with observations made in two previous studies—a national study of MTBE occurrence in ambient ground water and a regional study for the Northeast and Mid-Atlantic regions in drinking water. Although detections of gasoline compounds collectively and BTEX compounds were greater in source-water samples from areas with 10 or more gasoline storage tanks, MTBE detections were not significantly related to the density of gasoline storage tanks near the source waters. Similar observations regarding the lack of an association between MTBE detections in drinking water and storage-tank density and (or) proximity were made by previous investigations in California and Maine. This observation reflects the enhanced mobility and recalcitrance of MTBE relative to other gasoline contaminants and points to a greater variety of potential nonpoint MTBE sources such as small leaks and spills, urban runoff, recreational watercraft use, and atmospheric transport.

The occurrence of several subgroups of VOCs were shown to be statistically related to certain anthropogenic factors. The frequency of detection of any VOC, non-THM VOC, solvent, and refrigerant was significantly greater in areas where the percentage of urban land use exceeded 20 percent and where population density exceeded 1,000 people/mi2. Although the frequency of detection of any gasoline-related compound in general, and MTBE in particular, was not related to percentage of urban land use, the frequency of detection was greater in the high population-density areas and in areas with high MTBE use as an oxygenate in gasoline. In particular, MTBE detections were five times more frequent in source waters in high MTBE-use areas than outside of these areas. Conversely, BTEX detections were greater in more urbanized areas, but not directly related to population density or to high MTBE-use areas. Finally, the density of storage tanks around source waters affects the frequency of detection of gasoline-related VOCs in general, and BTEX in particular, but not MTBE.

FOCUSED SURVEY

The intent of the Focused Survey was to provide an improved understanding of the occurrence, concentration, temporal variability, and some anthropogenic factors associated with the frequent occurrence of MTBE and other frequently identified VOCs. A total of 451 samples were collected at fixed intervals for one year from 78 ground-water, 39 reservoir and (or) lake, and 17 river and (or) stream source waters in select areas of the nation. Ground-water source waters were sampled biannually, reservoirs and lakes were sampled quarterly, and rivers and streams were sampled eight times per year. In addition, a total of 744 field QC samples, including trip-reagent blanks, field-reagent blanks, field-equipment blanks, and duplicate samples were collected. All data underwent an exhaustive QC review process to accept or reject source-water data for inclusion in the Focused Survey's findings. The samples were analyzed for a total of 66 VOCs, including MTBE and 3 other gasoline oxygenates, at the MWDSC laboratory. Half of the samples were also analyzed for TBA, a gasoline additive as well as a degradation product of MTBE, at OHSU.

Occurrence data from the Focused Survey were compared to results obtained from the Random Survey using a common assessment level of 0.2 μg/L. A total of 42 individual VOCs were detected in each survey, 37 of which were detected in both surveys. However, the detection frequency of many VOCs was greater in samples collected as part of the Focused Survey than in the Random Survey. Because source waters suspected or known to contain MTBE were sampled in the Focused Survey, MTBE was detected in 55.5 percent of the 134 CWSs sampled in the Focused Survey, whereas MTBE was detected in only 8.7 percent of 954 CWSs sampled in the Random Survey. The increased detection frequency of other VOCs in the Focused Survey may be explained by high-population density, which was one factor associated with the selection of source waters to be sampled as part of the Focused Survey.

The 66 VOCs analyzed were subdivided into seven subgroups of VOCs, based mainly on the primary use of the compounds. Each subgroup of VOCs was found more frequently in the Focused Survey than in the Random Survey. The gasoline oxygenate subgroup, which includes MTBE, was found most frequently and was detected in 53.7 percent of CWSs in the Focused Survey and in 8.5 percent of CWSs sampled in the Random Survey. Again, this difference is due largely to the selection of sites with suspected or known occurrence of MTBE. A large difference in occurrence also was observed in the solvent subgroup; solvents were detected in 21.6 percent and 6.3 percent of CWSs sampled in the Focused and Random Survey, respectively.

The two most frequently detected solvents in both Surveys were tetrachloroethene and trichloroethene; however, these compounds were mainly detected only in CWSs supplied by ground water. Little difference in occurrence was observed for the THM subgroup; these compounds were detected in 16.4 percent of Focused Survey and 14.8 percent of the Random Survey CWSs, respectively. This may indicate a pervasive occurrence of THMs in source waters, regardless of CWS size or geographic location.

In contrast to results of the Random Survey, the occurrence of subgroups of VOCs does not appear to be related to the size (based on population served) of CWSs in the Focused Survey supplied by ground water. In the Focused Survey, each subgroup of VOCs was detected at similar frequencies in each CWS size category. However, there were some differences in CWSs supplied by surface water in that the gasoline oxygenate and other gasoline compound subgroups were predominantly detected in the Focused Survey, whereas solvents, THMs, and organic synthesis compounds were detected more frequently in several sizes of CWSs sampled in the Random Survey.

Only two samples were collected from each ground-water source water included in the Focused Survey; thus, temporal variations on a seasonal basis could not be identified. However, comparisons were made between the first sample and the second sample collected from 78 CWSs. No apparent differences were observed for any subgroup of VOCs between the first and second ground-water source water samples analyzed. It is presumed that an insufficient amount of time passed between the collection of the first and second ground-water sample, typically 6 months, to observe any significant changes in ground-water quality. The occurrence of each subgroup of VOCs was virtually identical between the two samples and no statistically significant differences at an alpha of 0.05, were noticed in the detected VOC concentrations. Detected concentrations were small, typically less than 1 μg/L. However, the maximum concentrations of MTBE, tetrachloroethene, and trichloroethene in ground-water source waters were 924, 290, and 165 μg/L, respectively.

On the basis of data at or above the method detection limit, concentrations of VOCs detected in the Focused Survey in reservoir and (or) lake source waters were typically small (less than 1 μg/L). Furthermore, very few statistically significant differences in detected concentrations were noticed on a seasonal basis among subgroups of VOCs. However, a slight seasonal pattern was evident for the occurrence of gasoline oxygenates and other gasoline compounds in the 39 reservoirs and lakes sampled. Both of these subgroups of VOCs were detected more frequently during spring and summer than in fall and winter. Thirty-three of the 39 reservoirs and (or) lakes sampled allow motorized watercraft. Thus, the greater occurrence of these compounds during spring and summer is presumed to be related to the increased use of watercraft during these seasons. No seasonal pattern was observed for the THMs or organic synthesis subgroups. This indicates the possibility of a common and continuous source of contamination for these VOCs.

As in ground-water and reservoir/lake source waters, concentrations of VOCs in river and (or) stream source waters typically were small (less than 1 μg/L), and very few statistically significant differences in detected concentrations were observed on a seasonal basis among subgroups of VOCs. Similarly, a seasonal pattern was not evident for the detection frequency of any subgroup of VOCs in the 17 river and (or) stream source waters sampled. Gasoline oxygenates, other gasoline compounds, THMs, and organic synthesis compounds were detected at similar frequencies within their respective subgroups year-round. The lack of seasonal differences for these subgroups of compounds may indicate a common and continuous source of contamination upstream of the intake location.

The most frequently detected VOC (MTBE) and the two most frequently occurring subgroups of VOCs (gasoline oxygenates and THMs) detected in CWS source waters sampled in the Focused Survey were further characterized to evaluate some anthropogenic factors that may better explain their frequent occurrence. MTBE, gasoline oxygenates, and THMs were detected in 77.3, 73.9, and 47.8 percent of CWS source waters sampled, respectively.

Fifty-five of 78 wells, 25 of 39 reservoirs/lakes, and 11 of 17 rivers/streams were inside high MTBE-use areas. Concentrations of MTBE in reservoir and (or) lake and river and (or) stream source waters inside and outside of high MTBE-use areas were also compared seasonally. MTBE was detected more frequently and concentrations were greater in high MTBE-use areas than in MTBE low-use areas. However, because as few as 3 detections of MTBE were found in surface-water source waters in low-MTBE-use areas, statistical comparisons were not completed.

The concentration of gasoline oxygenates in reservoirs was slightly correlated with watercraft use (boats/year/acre) on reservoirs inside high MTBE-use areas (r^2=0.3783). As watercraft use increased, the average concentration of gasoline oxygenates also increased. No relation between gasoline oxygenates in reservoirs and watercraft use was observed in reservoirs outside of high MTBE-use areas (r^2=0.0242).

Trihalomethanes were detected in 11 of the reservoirs/lakes sampled. Seven of these reservoirs contained detectable concentrations of more than one THM. Each of the 7 CWSs provided additional information on reasons why THMs may occur in their source water. In all cases, the presence of THMs were the result of the addition of chlorine to waters that ultimately recycled back to, or upstream of, the reservoir that was sampled. That is, the THMs were present in the source water as a result of recycling chlorinated water.

TBA was analyzed for about half of the samples collected in the Focused Survey. TBA was detected in 138 out of a total of 273 samples collected (51 percent). Seasonal trends were not observed and the median values ranged from 0.28 µg/L in the summer to 0.37 µg/L in the fall. There did not appear to be any difference in occurrence for the various ground or surface water types (median values ranged from 0.21 µg/l in lakes to 0.65 µg/L in wells. System size did show some interesting differences with the highest median value (3.72 µg/L) observed in the small systems. There was no observed pattern of occurrence of TBA relative to state or region. In this survey, a particular site was not sampled intensively enough to provide evidence for in situ degradation of MTBE to TBA, but an extensive survey such as this might provide insight into any casual relationship between MTBE and TBA occurrence via analysis of statistical correlations. A positive, but not very strong, correlation between TBA occurrence and MTBE occurrence was observed. This weak correlation might be due to differential dissolution and transport of TBA relative to MTBE and additional sources of TBA other than a degradation by-product of MTBE (e.g., use of TBA as an oxygenate). A more in depth investigation of TBA in source waters is warranted, based on these preliminary TBA findings and such a follow-up survey should be designed to determine the source of TBA--MTBE degradation, the atmosphere, and/or gasoline itself.

CHAPTER 7
RECOMMENDATIONS TO COMMUNITY WATER SYSTEMS

RECOMMENDATIONS

MTBE is used widely and is the fourth highest produced chemical in the United States. Its characteristics (high solubility in water, low soil-sorption properties, and recalcitrant nature in the environment) make water supplies susceptible to contamination. In general, high concentrations of VOCs are not thought to be persistent in surface waters due to volatilization to the atmosphere. However, high concentrations of VOCs may be found in ground water at point-source release sites. Literature surveys have identified MTBE in a large number of states. Although concentrations in drinking-water systems are usually low (i.e., generally below 20 μg/L), the frequency of occurrence is high. In this study, MTBE was found to be the second most frequently detected VOC (second to chloroform). Once again, although the concentrations were low, the frequency of detection (8.7 percent) is telling of an environmental problem. CWSs need to be aware of the increase vulnerability of their source waters to MTBE.

This study showed that MTBE was detected more frequently in surface-water than ground–water supplies, which is most likely related to the use of motorized watercraft on open reservoirs and lakes. The occurrence seemed to be more frequent in the warm weather months when there was an increase in motorized watercraft use. In the Focused Survey, there was a direct correlation between gasoline oxygenate concentration and the watercraft use factor (number of boats per acre per year) for reservoirs/lakes that were located in high MTBE-use areas. This indicates that CWSs that draw water from reservoirs, lakes, or rivers where motorized watercraft are allowed should maintain a monitoring program that characterizes the extent of the contamination. The frequency of monitoring should be increased during the warm weather months (April through September) to evaluate conditions during the most vulnerable season when watercraft are being used. CWS should consider limiting the number of watercraft or requiring the use of MTBE free fuel on the reservoirs and lakes to reduce the input of MTBE. Monitoring programs of lakes and reservoirs should include depth profiles to determine the dispersion of MTBE throughout the water column. Monitoring should include other oxygenates because of their potential use as a substitute for MTBE.

The frequency of detecting MTBE in source-waters was found to relate to areas where MTBE is used in high volume. High MTBE-use areas identified in this study were areas within the federal RFG program plus additional areas in the State of California and Yuma/Mojave counties in Arizona. In addition, population density and urban land use were also studied. It was found that the frequency of detection for any VOC was greater in more urbanized areas.

Similarly, high population-density areas were more likely to have MTBE and other gasoline components (e.g., benzene, toluene, ethylbenzene and xylenes) detected. MTBE was found to co-occur with some of the other gasoline components in only a few of the samples from the Random Survey. The other oxygenates were detected rarely (probably due to their limited use), and when detected, were found to co-occur with MTBE. These findings indicate that CWSs in high population-density areas (especially within high-MTBE-use areas) should maintain a routine monitoring program for MTBE, gasoline components, and other VOCs. The use of other gasoline oxygenates may increase in the future. Because of their similarity in chemical properties to MTBE it is recommended that these compounds should also be included in a routine monitoring program. These monitoring programs should include both ground- and surface-water systems.

Gasoline compounds (i.e., BTEX) were found to occur in correlation with chemical (fuel) storage tanks. However, the literature and this study did not find a correlation between the density of fuel storage tanks and the occurrence of MTBE. The lack of a relation between MTBE occurrence and fuel storage tanks should not reduce the concern of possible contamination or the need for an aggressive monitoring program because this observation may reflect the enhanced mobility, recalcitrance, and other sources of MTBE, relative to the other gasoline products.

There are a variety of MTBE contamination non-point sources that include small leaks and spills, urban runoff, recreational watercraft use and atmospheric deposition. MTBE was found in some areas outside high MTBE-use areas, which may be due to the small use of MTBE in gasoline for octane enhancement. The pattern of MTBE's use and these sources of contamination imply that areas outside high MTBE-use areas, especially high population-density areas, should also have a routine MTBE monitoring program in place.

THMs were found frequently in this study in both ground and surface water sources. The frequency of occurrence of THMs in this study was not expected because the sampling was designed to collect source (i.e., untreated) water. Chloroform was found to be the most frequently detected compound (13 percent). In the Focused Survey, the occurrence of THMs in some of the surface waters could have been the result of the addition of chlorine to water that ultimately recycled back to, or upstream of, the source water that was sampled. THMs detected in ground water may be due to past chlorination practices and eventual percolation of chlorinated water into the ground water aquifers. Chlorofom, however, could originate from other sources (manufacture of pharmaceuticals, dry cleaning, fire extinguishers, and fumigants) than from by-products of chlorination. The frequency of occurrence of THMs was independent of CWS size. There was no seasonal trend observed for the occurrence of THMs, which indicates a possible common and continuous source of contamination. These factors indicate that CWSs (of all sizes) should include THMs as part of the their routine source-water monitoring programs. The surprisingly frequent occurrence of THMs, and chloroform in particular, may imply a source of contamination and source water monitoring should be designed to observe any trends.

Solvents were the third most frequently detected subgroup of VOCs in this study. The occurrence of solvents primarily were found in ground water. The concentrations of some solvents in some of the CWSs were higher than MTBE and the THMs. Overall, the occurrence of the solvents was typically at concentrations below MCLs and no seasonal effects were observed. However, since the use of solvents is widespread, have known human-health concerns, and can be analyzed in the same analytical method as MTBE and the other gasoline compounds, solvents should also be included in a routine source-water monitoring program. The solvents most frequently detected were tetrachloroethene, trichloroethene, cis-1,2-dichloroethene, 1,1-dichloroethane, 1,2-dichloroethane, and 1,1,1-trichloroethane.

Analytical data indicate that there is a positive correlation between the detection of TBA and MTBE. However, TBA was not studied extensively enough in this project to determine the in-situ degradation of MTBE. It is recommended that CWSs in high MTBE-use areas and in high population-density areas include TBA in their routine source-water monitoring programs. Other locations where MTBE has been detected should also monitor for TBA.

The field quality-control samples collected in this study demonstrated that contamination of VOC samples may occur during the collection of these samples and/or during sample shipment. This finding demonstrates the importance of training for water utility operators in proper sample collection and field handling, as well as the importance of including field blanks in monitoring programs.

In summary, CWSs should establish a routine monitoring program of their respective source waters for MTBE, THMs, gasoline compounds, solvents, other oxyenates, and TBA. The monitoring program in surface water should include temporal monitoring to observe seasonal effects and depth profiling in reservoirs and lakes to observe dispersion. Ground water systems should monitor routinely, but short-term seasonal changes are not expected (unless the ground water is under the direct influence of surface water). CWSs within high-MTBE-use areas and CWSs within high population-density areas especially should establish routine source-water monitoring programs.

REFERENCES

Alabama Department of Environmental Management, Water Division. 2001. Evaluation of MTBE—Contamination of Drinking Water Sources in Alabama. Accessed June 8, 2001, at URL http://www.adem.state.al.us/ EduInfo/PressReleases/3dwsafe.htm

Alford-Stevens, A., J.W. Eichelberger, and W.L. Budde. 1983. Purgeable Organic Compounds in Water by Gas Chromatography/Mass Spectrometry, Method 524. Environmental Monitoring and Support Laboratory, U.S. Environmental Protection Agency, Cincinnati, Ohio.

ACS (American Chemical Society). 1979. Safety in Academic Chemistry Laboratories. American Chemical Society Publication, Committee on Chemical Safety, 3rd ed. Washington, D.C.: ACS.

ASTM (American Society for Testing and Materials). 1991. E 679-91, Standard Practice for Determination of Odor and Taste Thresholds by a Forced-Choice Ascending Concentration Series Method of Limits, ASTM volume 15.07. Philadelphia, Pa.: ASTM

API (American Petroleum Institute). 1994. Odor Threshold Studies Performed with Gasoline and Gasoline Combined with MTBE, ETBE, and TAME (conducted by TRC Environmental Corporation). Washington, D.C.: API.

ARCO Chemical. 1993. The Odor and Taste Threshold Studies Performed with Methyl Tertiary-Butyl Ether and Ethyl Tertiary-Butyl Ether (conducted by TRV Environmental Corporation). Houston, Texas: ARCO.

Arizona Department of Environmental Quality. 1999. Report on Methyl Tertiary-Butyl Ether (MTBE). Accessed March 1, 2002, at URL http://www.adeq.state.az.us/environ/waste/ust/mtbe.html.

Baehr, A.L., P.E. Stackelberg, and R.J. Baker. 1999. Evaluation of the Atmosphere as a Source of Volatile Organic Compounds in Shallow Groundwater. *Water Resources Research*, 35(1):127-136.

Barreto, R.D., K.A. Gray, and K. Anders. 1995. Photocatalytic Degradation of Methyl tert-Butyl Ether in TiO2 Slurries: A Proposed Reaction Scheme. *Wat. Res.*, 29:1243-1248.

Bender, D.A., J.S. Zogorski, W. Luo, J.F. Pankow, M.S. Majewski, and R.J. Baker. 2000. Atmospheric-Water Interaction of Chloroform, Toluene, and MTBE in Small Perennial Urban Streams. In *Proc. of the 93rd Air and Waste Management Association Conference and Exposition*. Salt Lake City, Utah: AWMA.

Boughton, C.J., and M.S. Lico. 1998. Volatile Organic Compounds in Lake Tahoe, Nevada and California, July-September 1997. U.S. Geological Survey Fact Sheet FS-055-98.

Bourelle, A. 1998. MTBE--A Real Concern for South Lake Tahoe. Tahoe Daily Tribune, August 28, 1998. Accessed Dec. 6, 2000, at URL http://www.tahoe.com/tribune/stories.8.28.98/news/mt28Aug1347.html

Bush, P.W., A.F. Ardis, L. Fahlquist, P.B. Ging, C.E. Horning, and J. Lanning-Rush. 2000. Water Quality in South-Central Texas, 1996-98. U.S. Geological Survey Circular 1212:32.

California Environmental Protection Agency. 1997. MTBE (Methyl Tertiary Butyl Ether). Briefing paper prepared by the California Environmental Protection Agency, April 24, 1997 (updated September 3, 1998). Accessed February 28, 2002 at URL http://www.arb.ca.gov/cbg/oxy/oxy.htm.

California Senate. 1998. Does California Need MTBE? Accessed February 28, 2002 at URL http://www.sen.ca.gov/ftp/sen/sor/environ/98mtbe.htm.

California Department of Health Services. 2001. MTBE in California Drinking Water. Accessed Dec. 6, 2000, at URL http://www.dhs.cahwnet.gov/ps/ddwem/chemicals/MTBE/mtbeindex.htm

California Department of Health Services. MTBE in California Drinking Water. Drinking water program download page. Accessed March 26, 2002, at URL http://www.dhs.ca.gov/ps/ddwem/publications/

Church, C. D., J.F. Pankow, and P.G. Tratnyek. 1999. Hydrolysis of tert-Butyl Formate: Kinetics, Products, and Implications for the Environmental Impact of MTBE. *Environ. Toxicol. Chem.*, 18:2789-2796.

Church, C. D., P.G. Tratnyek, and K.M. Scow. 2000. Pathways for the Degradation of MTBE and Other Fuel Oxygenates by Isolate PM1. In *Proc. of the 219th ACS National Meeting, Division of Environmental Chemistry*, 40(1):261-263. San Fransisco, Calif.: ACS.

City of Santa Monica. 1999. MTBE issue featured on "60 Minutes." Accessed Dec. 6, 2000, at URL http://epwm.santamonica.org/epwm/watquality/1999report/waterweb/news05.htm

Connor, B. F., D.L. Rose, M.C. Noriega, L.K. Murtagh, and S.R. Abney. 1998. Methods of Analysis by the U.S. Geological Survey National Water Quality Laboratory—Determination of 86 Volatile Organic Compounds in Water by Gas Chromatography/Mass Spectrometry, Including Detections Less Than Reporting Limits. U.S. Geological Survey Open-File Report 97-829.

Dale, M.S., M.S. Moylan, B. Koch, and M.K. Davis. 1997. MTBE: Taste and Odor Threshold Determinations Using the Flavor Profile Method. In *Proc. of the AWWA Water Quality Technology Conference*. Denver, Colo.: AWWA.

Dale, M.S., B. Koch, R.F. Losee, E.W. Crofts, and M.K. Davis. 2000. MTBE in Southern California Water—Motorized Recreation Watercraft Were the Predominant Mode of Contamination in Reservoirs. *Jour. AWWA*, 92(8):42-51.

Delzer, G.C., J.S. Zogorski, T.J. Lopes, and R.L. Bosshart. 1996. Occurrence of the Gasoline Oxygenate MTBE and BTEX Compounds in Urban Stormwater in the United States, 1991- 95. U.S. Geological Survey Water-Resources Investigations Report 96-4145:6.

Eichelberger, J. W., J.W. Munch, and T.A. Bellar. 1992. Method 524.2— Measurement of Purgeable Organic Compounds in Water by Capillary Column Gas Chromatography/Mass Spectrometry, Revision 4.0. Cincinnati, Ohio: EPA.

Florida Department of Environmental Protection. 2001. MTBE (Methyl Tertiary-Butyl Ether) Occurrence in Florida Drinking Water Systems. Accessed June 8, 2001, at URL http://www.dep.state.fl.us/water/drinkingwater/mtbe.htm

Gabele, P.A., and S.M. Pyle. 2000. Emissions From Two Outboard Engines Operating on Reformulated Gasoline Containing MTBE. *Environ. Sc.e & Technol.*, 34(3):368-372.

Gilliom, R.J., W.M. Alley, and M.E. Gurtz. 1995. Design of the National Water-Quality Assessment Program— Occurrence and Distribution of Water-Quality Conditions. U.S. Geological Survey Circular 1112:33.

Glaser, J.A., D.L. Foerst, G.D. McKee, S.A. Quave, and W.L. Budde. 1981. Trace Analyses for Wastewaters. *Environ. Sci. Technol.*, 15:1426.

Gomez, J., T. Brasil, and N. Chan. 1998. An Overview of the use of Oxygenates in Gasoline, State of California. Accessed at URL http://www.arb.ca.gov/cbg/pub/oxyrpt.pdf

Grady, S.J. 1994. Effects of Land Use on Quality of Water in Stratified-Drift Aquifers in Connecticut—Chapter B, Analysis of Nonpoint-Source Ground-Water Contamination in Relation to Land Use. U.S. Geological Survey Water-Supply Paper 2381:56.

Grady, S.J., and J.R. Mullaney. 1998. Natural and Human Factors Affecting Shallow Water Quality in Surficial Aquifers in the Connecticut, Housatonic, and Thames River Basins. U.S. Geological Water-Resources Investigations Report 98-4042:81.

Grady, S.J., and G.D. Casey. 2001. Occurrence and Distribution of Methyl *tert*-Butyl Ether and Other Volatile Organic Compounds in Drinking Water in the Northeast and Mid-Atlantic Regions of the United States, 1993-98. U.S. Geological Survey Water-Resources Investigations Report 00-4228:123.

Grady, S.J. In press. A National Survey of Methyl tert-Butyl Ether (MTBE) and Other Volatile Organic Compounds in Drinking Water Sources: Results of the Random Source-Water Survey. U.S. Geological Survey Water-Resources Investigations Report 02-4079.

Gullick, R.W., and M.W. LeChevallier. 2000. Occurrence of MTBE in Drinking Water Sources. *Jour. AWWA*, 92(1):100-113.

Halde, M.J., G.C. Delzer, and J.S. Zogorski. 1998. Study Design and Analytical Results Used to Evaluate a Surface-Water Point Sampler for Volatile Organic Compounds. U.S. Geological Survey Open-File Report 98-651:31.

Hanson, J. R., C.E. Ackerman, and K.M. Scow. 1999. Biodegradation of Methyl *tert*-Butyl Ether by a Bacterial Pure Culture. *Appl. Environ. Microbiol.*, 65:4788-4792.

Happel, A.M., B. Dooher, and E.H. Beckenbach. 1998. Methyl Tertiary Butyl Ether (MTBE) Impacts to California Groundwater. Lawrence Livermore National Laboratory, URCL-AR-130897:68. Livermore, Calif.: LLNL.

Hardison, L. K., S.S. Curry, L.M. Ciuffetti, and M.R. Hyman. 1997. Metabolism of Diethyl Ether and Cometabolism of Methyl tert-Butyl Ether by a Filamentous Fungus, a Graphium sp. *Appl. Environ. Microbiol.*, 63:3059-3076.

Hattan, G. 2000. MTBE Treatment Systems in Kansas. Association of State and Territorial Solid Waste Management Officials, MTBE and Fuel Oxygenates Workgroup Newsletter, 3(1):31.

Helsel, D.R., and Hirsch, R.M, 1992. *Statistical Methods in Water Resources*. New York: Elsevier Science.

Hitzig, R., P. Kostecki, and D. Leonard. 1998. Study Reports LUST Programs are Feeling Effects of MTBE Releases. *Soil and Groundwater Cleanup*, August/September, p. 15-19.

Hunter, B. 1999. MTBE & other Toxics in Maine's Environment: Impacts of Small Gasoline Spills on Groundwater. In *Proc. of the 1999 Maine Water Conference*. Augusta, Maine: Maine Department of Environmental Protection.

Illinois Environmental Protection Agenc. 2000. MTBE—Good News/Bad News Additive, Good News for Air Quality is Not-So-Good News for Groundwater. Accessed on April 10, 2001, at URL www.epa.state.il.us/environmental-progress/v25/nl/ mtbe.html

Iman, R.L., and W.J. Conover. 1983. *A Modern Approach to Statistics*. New York:John Wiley and Sons.

Iowa Department of Natural Resources. 2000. Methyl Tertiary-Butyl Ether (MTBE) Occurrence in Iowa. Presented at the 2000 Session of the 78th General Assembly, Iowa Department of Natural Resources, Underground Storage Tank Section, Des Moines, Iowa.

Ivahnenko, T., S.J. Grady, and G.C. Delzer. 2001. Design of a National Assessment of MTBE and Other Volatile Organic Compounds in Sources of Drinking Water. U.S. Geological Survey Open-File Report 01-271:36.

Johnson, R.L., J.F. Pankow, D.A. Bender, C.V. Price, and J.S. Zogorski. 2000. MTBE—To What Extent Will Past Releases Contaminate Community Water Supply Wells? *Environ. Sci. and Technol.*, 34(9):2A-9A.

Juttner, F., D. Backhaus, U. Matthias, U. Essers, R. Greiner, and B. Mahr. 1995. Emissions of Two- and Four-Stroke Outboard Engines; Quantification of Gases and VOC. *Wat. Res.*, 29(8):1976-1982.

Kang, J. W., and M.R. Hoffmann. 1998. Kinetics and Mechanism of the Sonolytic Destruction of Methyl tert-Butyl Ether by Ultrasonic Irradiation in the Presence of Ozone. *Environ. Sci. Technol.*, 32:3194-3199.

Kansas Department of Health and Environment. 2000. Testimony Presented to Senate Committee on Energy and Natural Resources, Division of Environment, Topeka, Kan.

Keller, A.K., J. Froines, C. Koshland, J. Reuter, I.H. Suffet, and J. Last. 1998. Health and Environmental Assessment of MTBE, Vol. I - Summary and Recommendations. Report to the Governor and Legislature of the State of California as Sponsored by SB 521.

Koch, B., 2002 Written communication to USGS regarding carry over of naphthalene from analysis to analysis.

Koterba, M.T., F.D. Wilde, and W.W. Lapham. 1995. Ground-Water Data-Collection Protocols and Procedures for the National Water-Quality Assessment Program— Collection and Documentation of Water-Quality Samples and Related Data. U.S. Geological Survey Open-File Report 95-399:113.

Landmeyer, J.E., F.H. Chapelle, P.M. Bradley, J.F. Pankow, C.D. Church, and P.G.Tratnyek. 1998. Fate of MTBE Relative to Benzene in a Gasoline-Contaminated Aquifer (1993-98). *Ground Water Monit. Rem.*, 18(4): 93-102.

Lopes, T.J., and D.A. Bender. 1998. Nonpoint Sources of Volatile Organic Compounds in Urban Areas—Relative Importance of Land Surfaces and Air. *Environmental Pollution*, 101:221-230.

Lopes, T.J., and E.T. Furlong. 2001. Occurrence and Potential Adverse Effects of Semivolatile Organic Compounds in Streambed Sediment, United States, 1992-1995. *Environmental Toxicology and Chemistry*, 20(4):727-737.

Luo, W., L.M. Isabelle, and J.F. Pankow. 2001. Determination of Volatile Organic Compounds in Water by Purge and Trap Gas Chromatography/Mass Spectrometry. Research report, Oregon Health & Science University, Portland, Ore.

Madding, C. 1984. Volatile Organic Compounds in Water by Purge and Trap Capillary Column GC/MS. In *Proc. of thel AWWA Water Quality Technology Conference*. Denver, Colo.: AWWA.

Maine Bureau of Health. 1998. The Presence of MTBE and Other Gasoline Compounds in Maine's Drinking Water. Preliminary report. Auburn, Maine: State of Maine, Department of Human Services.

Maryland Department of the Environment. 2001. MTBE Update. Accessed June 8, 2001, at URL http:// www.mde.state.md.us/was/mtbe.htm

Michigan Department of Environmental Quality, Storage Tank Division. 2000. MTBE (Methyl Tertiary Butyl Ether) Fact Sheet. Accessed June 12, 2001, at URL http://www.deq.state.mi.us/std/MTBE.pdf

Missouri Department of Natural Resources. 2000. Report on Inter-Agency Meetings on MTBE. Accessed October 2, 2001, at URL http://www.pstif.org/apps/mtbe_report.pdf

Minear, R.A., and G.L. Amy, eds. 1995. *Disinfection By-Products in Water Treatment—The Chemistry of Their Formation and Control*. New York: Lewis Publishers.

Moran, M.J., M.J. Halde, R.M. Clawges, and J.S. Zogorski. 1999a. Relations Between the Detection of Methyl *tert*-Butyl Ether (MTBE) in Surface and Ground Water and its Content in Gasoline. In *Proc. of the 219th American Chemical Society National Meeting*. San Francisco, Calif.,: ACS.

Moran, M.J., J.S. Zogorski, and P.J. Squillace. 1999b. MTBE in Ground Water of the United States—Occurrence, Potential Sources, and Long-Range Transport, In *Proc. of the AWWA Water Resources Conference*. Denver, Colo.: AWWA.

Moran, M.J., W.W. Lapham, B.L. Rowe, and J.S. Zogorski. 2002a. Occurrence and Status of Volatile Organic Compounds in Ground Water From Rural, Untreated, Self-Supplied Domestic Wells in the United States, 1986-99. U.S. Geological Survey Water-Resources Investigations Report 02-4085.

Moran, M.J., R.M. Clawges, and J.S. Zogorski. 2002b Methyl *tert*-Butyl Ether in Ground and Surface Water of the United States. In *Oxygenates in Gasoline—Environmental Aspects*. Edited by A. L. Diaz and D. L. Drogos. Washington, D.C: ACS.

Munch, J. W. 1995. Method 524.2—Measurement of Purgeable Organic Compounds in Water by Capillary Column Gas Chromatography/Mass Spectrometry, Revision 4.1. Cincinnati, Ohio: EPA.

National Water Research Institute. 1999. Treatment Technologies for Removal of Methyl Tertiary Butyl Ether (MTBE) form Drinking Water: Air Stripping, Advanced Oxidation Processes, Granular Activated Carbon, Synthetic Resin Sorbents, Second Edition. Project No. NWRI-99-06 http://www.nwri-usa.org/asp/sp.asp?main=m4&sub=s1&id=4&sub2=4

New Jersey Department of Environmental Protection. 1999. MTBE in New Jersey's Environment. Accessed October 2, 2001, at URL http://www..state.nj/us/dep/dsr/mtbe/mtbe-conclusion.htm

Occupational Safety and Health Administration. 1976. OSHA Safety and Health Standards, General Industry. (29CFR1910), , OSHA 2206, (Revised, January 1976). Washington, D.C.: OSHA.

O'Reilly, K. T., M.E. Moir, C.D. Taylor, C.A. Smith, and M.R. Hyman. 2001. Hydrolysis of tert-Butyl Methyl Ether (MTBE) in Dilute Aqueous Acid. *Environ. Sci. Technol.*, 35:3954-3961.

Pankow, J.F., N.R. Thomson, R.L. Johnson, A.L. Baehr, and J.S. Zogorski. 1997. The Urban Atmosphere as a Non-Point Source for the Transport of MTBE and Other Volatile Organic Compounds (VOCs) to Shallow Ground Water. *Environ. Sci. & Technol.*, 31(10):2821-2828.

Pankow, J. F., W. Luo, L.M. Isabelle, D.A. Bender, and R.J. Baker. 1998. Determination of a Wide Range of Volatile Organic Compounds in Ambient Air Using Multisorbent Adsorption/Thermal Desorption and Gas Chromatography/Mass Spectrometry. *Anal. Chem.*, 70(24):5213-5221.

Pomes, M.L., W.R. Green, E.M. Thurman, W.H. Orem, and H.E. Lerch. 1999. DBP Formation Potential of Aquatic Humic Substances. *Jour. AWWA*, 91(3):101-115.

Price, C.V., and R.M. Clawges. 1999. Digital Data Sets Describing Water Use, Toxic Chemical Releases, Metropolitan Areas, and Population Density of the Conterminous United States. U.S. Geological Survey Open-File Report 99-78, scale 1:2,000,000 [CD-ROM disk].

Reuter, J.E., B.C. Allen, R.C. Richards, J.F. Pankow, C.R. Goldman, R.L. Sholl, and J.S. Seyfried. 1998. Concentrations, Sources, and Fate of the Gasoline Oxygenate Methyl *tert*-Butyl Ether (MTBE) in a Multiple-Use Lake. *Environ. Sci. & Technol.*, 21(23):3666-3672.

Reiser, E.G., and A.K. O'Brien. 1989. Occurrence and Seasonal Variability of Volatile Organic Compounds in Seven New Jersey Streams. U.S. Geological Survey Water-Resources Investigations Report 98-4074:11.

Robbins, G.A., B.J. Henebry, B.M. Schmitt, F.B. Bartolomeo, A. Green, and P. Zack. 1999. Evidence for MTBE in Heating Oil. *Ground Water Monitoring Review*, 19(2):65-69.

Rowe, B.L., S.J. Grady, J.S. Zogorski, B. Koch, and P.G. Tratnyek. 2001. Initial findings—National Survey of MTBE, Other Ether Oxygenates, and Other VOCs in Community Drinking-Water Sources. U.S. Geological Survey Open-File Report 01-399:1.

Salanitro, J. P., L.A. Diaz, M.P. Williams, and H.L. Wisniewski. 1994. Isolation of a Bacterial Culture That Degrades Methyl t-Butyl Ether. *Appl. Environ. Microbiol.*, 60:2593-2596.

Salanitro, J. P., P.C. Johnson, G.E. Spinnler, P.M. Maner, H.L. Wisniewski, and C. Bruce. 2000. Field Scale Demonstration of Enhanced MTBE Bioremediation Through Aquifer Bioaugmentation and Oxygenation. *Environ. Sci. Technol.*, 34(19):4152-4162.

Shelton, L.R. 1997. Field Guide for Collecting Samples for Analysis of Volatile Organic Compounds in Stream Water of the National Water-Quality Assessment Program. U.S. Geological Survey Open-File Report 97-401:14.

Shelton, J.L., K.R. Burrow, K. Belitz, N.M. Dubrovsky, M. Land, and J. Gronberg. 2001. Low-level Volatile Organic Compounds in Active Public Supply Wells as Ground-Water Tracers in the Los Angeles Physiographic Basin, California, 2000. U.S. Geological Survey Water-Resources Investigations Report 01-4188:29.

Shen, Y.F., L.J. Loo, S.R. Fitzsimmons, M.K. Yamamoto. 1997. Threshold Odor Concentrations of MTBE and other Fuel Oxygenates. Presented at the American Chemical Society National Meeting, San Francisco, CA April 1997.

Squillace, P.J., J.F. Pankow, N.E. Korte, and J.S. Zogorski. 1997. Review of the Environmental Behavior and Fate of Methyl *tert*-Butyl Ether. *Environ. Toxicol. Chem.* 16(9):1836-1844.

Squillace, P.J., M.J. Moran, W.W. Lapham, C.V. Price, R.M. Clawges, and J.S. Zogorski. 1999. Volatile Organic Compounds in Untreated Ambient Groundwater of the United States, 1985-1995. *Environ. Sci. & Technol.*, 33(23):4176-4187.

Squillace, P.J., and M.J. Moran. 2000. Estimating the Likelihood of MTBE Occurrence in Drinking Water Supplied by Ground-Water Sources in the Northeast and Mid-Atlantic Regions of the United States. U.S. Geological Survey Open-File Report 00-343:10.

State of Connecticut Department of Public Health. 2000. Report per Section 25-39a of the Connecticut General Statutes: Monitoring, Health Risks, and Treatment Costs for Organic Chemicals in Public Drinking Water for the Year 1999. Hartford, Conn.: State of Connecticut, 25 p.

State of Maine. 1998. The Presence of MTBE and Other Gasoline Compounds in Maine's Drinking Water. Preliminary Report, October 13, State of Maine, Bureau of Health, Bureau of Waste Management and Remediation, Maine Geological Survey. Augusta, Maine: State of Maine.

Steffan, R. J., K. McClay, S. Vainberg, C.W. Condee, and D. Zhang. 1997. Biodegradation of the Gasoline Oxygenates Methyl tert-Butyl Ether, Ethyl tert-Butyl Ether, and tert-Amyl Methyl Ether by Propane-Oxidizing Bacteria. *Appl. Environ. Microbiol.*, 63:4216-4222.

Stocking, A.J., I.H. Suffet, M.J. McGuire, and M.C. Kavanaugh. 2001. Implications of an MTBE Odor Study for Setting Drinking Water Standards. *Jour. AWWA*, 93(3):95-105.

Symons, J.M., L.C. Bradley, Jr., and T.C. Cleveland. 2000. *The Drinking Water Dictionary.* Denver, Colo.: AWWA.

U.S. Environmental Protection Agency. 1992a. Guidelines Establishing Test Procedures for the Analysis of Pollutants (Part 136, Appendix B. Definition and Procedure for the Determination of the Method Detection Limit- Revision 1.11). U.S. Code of Federal Regulations, Title 40, revised as of July 1992, p.565-567.

U.S. Environmental Protection Agency. 1992b. Methods for the Determination of Organic Compounds in Drinking Water—Supplement II. Office of Research and Development, Environmental Monitoring Systems Laboratory, Cincinnati, Ohio, EPA-600/R-92-129.

U.S. Environmental Protection Agency. 1993. Technical and Economic Capacity of States and Public Water Systems to Implement Drinking-Water Regulations. Report to Congress, September, 1993. U.S. Environmental Protection Agency, G-134, 127 p., 2 app.

U.S. Environmental Protection Agency. 1996. Drinking Water Regulations and Health Advisories. Washington D.C., Office of Water, USEPA 822-R-96-001, 16 p.

U.S. Environmental Protection Agency. 1997. Drinking Water Advisory: Consumer Acceptability Advice and Health Effects Analysis on Methyl Tertiary-Butyl Ether (MTBE). U.S. Environmental Protection Agency, Office of Water, EPA-822-F-97-009:4.

U.S. Environmental Protection Agency. 1998a. Announcement of the Dinking Water Contaminant Candidate List, Notice (March 12, 1998). *Fed. Reg.*, 63(40):10273-10287.

U.S. Environmental Protection Agency. 1998b. Safe Drinking Water Information System. Accessed on November 5, 1998 and Dec. 12, 2000, at URL http://www.epa.gov/enviro/html/sdwis/

U.S. Environmental Protection Agency. 1999a. List of Reformulated Gasoline Program Areas. Office of Mobile Sources. Accessed April 5, 1999, at URL http://www.epa.gov/oms/rfgarea.htm.

U.S. Environmental Protection Agency. 1999b. Winter Oxygenated Fuels Program. Office of Mobile Sources. Accessed June 16, 1999, at URL http://www.epa.gov/oms/fuels.htm

U.S. Environmental Protection Agency. 1999c. Revisions to the Unregulated Contaminant Monitoring Regulation for Public Water Systems; Final rule (September 17, 1999). *Fed. Reg.*, 64(180):50556-50620.

U.S. Environmental Protection Agency. 2000a. List of Reformulated Gasoline Program Areas. Office of Transportation and Air Quality. Accessed Jan. 5, 2001, at URL http://www.epa.gov/omswww/rfgarea.htm

U.S. Environmental Protection Agency. 2000b. Drinking Water Standards and Health Advisories. Office of Water, EPA 822-B-00-001:17.

U.S. Environmental Protection Agency. 2001. National Primary Drinking Water Standards: Office of Water, EPA 816-F-01-007:4.
Vista Information Solutions. 1999. StarView Real Estate Version 2.6.1, Users Guide. San Diego, Calif.,: Vista Information Solutions.
Vogelmann, J.E., S.M. Howard, L. Yang, C.R. Larson, B.K. Wylie, and J.N. Van Driel. 2001. Completion of the 1990's National Land Cover Data Set for the Conterminous United States. *Photogrammetric Engineering and Remote Sensing*, 67:650-662.
Westrick, J.J. 1990. National Surveys of Volatile Organic Compounds in Ground and Surface Waters. In *Significance and Treatment of Volatile Organic Compounds in Water Supplies*. Edited by N.M. Ram, R.F. Christman, and K.P. Cantor, K.P. Chelsea, Mich.: Lewis Publishers.
Wigglesworth, T. 1999. MTBE—a Watertight Case. *Chemistry and Industry*, 10:408.
Wisconsin Department of Natural Resources Bureau for Remediation and Redevelopment. 2000. MTBE in Wisconsin Groundwater. *RE News*, 10(2):1-3.
Young, W. F., H. Horth, R. Crane, T. Ogden, and M. Arnott. 1995. Taste and Odour Threshold Concentrations of Potential Potable Water Contaminants. *Wat. Res.*, 30(2):331-340.
Zapecza, O.S., and Baehr, A.L., 1999. Methyl tert-Butyl Ether (MTBE) in Lakes in Byram Township, Sussex County, New Jersey, 1998, and Implications for Ground Water Vulnerability. In U.S. Geological Survey Toxics Hydrology Program—Proceedings of the Technical Meeting, Volume 2, Contamination of Hydrologic Systems and Related Ecosystems. D.W. Morganwalp, and H.T. Buxton, eds. Charleston, S.C., March 8-12, 1999 [Proceedings]: U.S. Geological Survey Water-Resources Investigations Report 99-4018B:405-414.
Zogorski, J.S., Morduchowitz, A.M., Baehr, A.L., Bauman, B.J., Conrad, D.L., Drew, R.T., Korte, N.E., Lapham, W.W., Pankow, J.F., and Washington, E.R., 1997. Fuel Oxygenates and Water Quality. In *Interagency Assessment of Oxygenated Fuels*. Washington, D.C.: National Science and Technology Council, Office of Science and Technology Policy.

GLOSSARY
Water Words (Symons 2000)

Ambient Surroundings. For example, a test performed at ambient temperature would be performed at the temperature surrounding the apparatus, i.e., laboratory temperature if the equipment were in a laboratory.

Ambient water quality standards Provisions of state or federal law set under the authority of the Clean Water Act that consist of a designated use or uses for the waters within the state or of the United States and water quality criteria for these waters based on the designated use. Ambient water quality standards protect public health and welfare, enhance the quality of surface waters, and serve the purpose of the Clean Water Act.

Appurtenances Machinery, appliances, structures, and other parts of a main structure that are necessary to allow it to operate as intended but are not considered part of the main structure.

Aquifer A geologic formation, group of formations, or part of a formation that is saturated and sufficiently permeable to transmit economic quantities of water to wells and springs.

Community water system (CWS) A public water system serving at least 15 service connections used by year-round residents or regularly serving at least 25 year-round residents. *See also* public water system.

Consumptive waste The water that returns to the atmosphere without benefiting people.

Contamination Any introduction into water of microorganisms, chemicals, wastes, or wastewater in concentrations that make the water unfit for its intended use.

Distribution sample A sample of water taken from the distribution piping or a site served by the distribution system.

Distribution system A system of conduits (laterals, distributaries, pipes, and their appurtenances) by which a primary water supply is distributed to consumers. The term applies particularly to the network of pipelines in the streets in a domestic water system.

Domestic use Water use by the public (home use).

Drinking water Water that meets or exceeds all applicable federal, state, and local requirements concerning safety. Drinking water is also called potable water. *See also terms under* water.

Drinking water regulation A regulation set by a state or by the US Environmental Protection Agency that applies to public water systems.

Drinking water standard (DWS) A numerical limit established by a state or federal agency for contaminants that may be found in drinking water.

Drinking water treatment plant The central facility for the production of potable water, containing all treatment processes and appurtenances exclusive of the distribution system.

Evapotranspiration (ET) (1) The process of water vapor entering the atmosphere as a result of both evaporation and transpiration. (2) Water vapor given off by plants through their leaves.

Finished water (product water) Water that has passed through a water treatment plant, such that all the treatment processes are completed, or finished. This water is ready to be delivered to consumers. It is also called product water.

Groundwater The water contained in interconnected pores located (1) below the water table in an unconfined aquifer or (2) in a confined aquifer.

Noncommunity water system (NCWS) A public water system that is not a community water system.

Nontransient, noncommunity water system (NTNCWS) A public water system that is not a community water system but that regularly serves at least 25 of the same people at least 6 months of the year. Examples of such systems include schools, factories, and hospitals that have their own water supplies. *See also* community water system; public water system; transient, noncommunity water system.

Palatability The degree to which water is significantly free of color, turbidity, taste, and odor, is of moderate temperature in summer and winter, and is well-aerated.

Palatable Agreeable or pleasant, especially to the sense of taste. *See also* palatability.

Potable water Water that is safe and satisfactory for drinking and cooking.

Private water Streams or other bodies of surface water that are not publicly owned. *See also* public water.

Private water supply A water supply not available to the general public because it is located on or has outlets on private property to which the public does not have access or legal right of entry.

Public water Water that is open to the public recreational use.

Public water supply *See* public water system.

Public water system (PWS) As defined in Section 1401 (4) of the Safe Drinking Water Act, as a system for providing to the public water for human consumption through pipes or other constructed conveyances, if such system has at least 15 service connections or regularly serves at least 25 individuals daily at least 60 days out of the year. Such a system includes (1) any collection, treatment, storage, and distribution facilities under control of the operator system and used primarily in connection with such system, and (2) any collection of pretreatment storage facilities not under such control but that are used primarily in connection with such system. A public water supply system is either a community water system or a noncommunity water system. On Aug. 5 1998, the US Environmental Protection Agency published guidelines regarding interpretation of the definition of *public water system* (*Federal Register,* 63:41939-41946). *See also* community water system; noncommunity water system; transient noncommunity water system.

Public water use The use of water to support public facilities, such as public buildings and offices, parks, golf courses, picnic grounds, campgrounds, and ornamental fountains

Raw water Water from the supply source prior to treatment. *See also* source water.

Safe drinking water Water that does not contain harmful bacteria or toxic materials or chemicals. Water may have taste-and-odor problems, color, and certain mineral problems and still be considered safe for drinking. *See also* potable water.

Source water (preferred over 'raw water') The supply of water for a water utility. Source water is usually treated before distribution to consumers, but some groundwaters are of such a quality that they can be distributed untreated.

Source Water Assessment Program (SWAP) A US Environmental Protection Agency program established by the 1996 Safe Drinking Water Act amendments, requiring states to develop a comprehensive program to identify the areas that supply public tap water, inventory contaminants and assess water system susceptibility to contamination, and inform the public of the results.

Surface water All water on the surface, as distinguished from subsurface water or groundwater.

Transient water system (TWS) A public water system that serves a transient population.

Treatment plant The central portion of water facilities that contains various treatment processes exclusive of collection or distribution of water.

Wastewater The used water and water-carried solids from a community (including used water from industrial processes) that flow to a treatment plant. Stormwater, surface water, and groundwater infiltration may also be included in the wastewater that enters a wastewater treatment plant. The term *sewage* usually refers to household wastes, but is has been replaced by the term *wastewater.*

Water (1) A transparent, odorless, tasteless compound of hydrogen and oxygen, H_20. At a pressure of 1 atmosphere (101.3 kPa), water freezes at 32° Fahrenheit (0° Celsius) and boils at 212° Fahrenheit (100°) Celsius. Water, in a more or less impure state, constitutes rain, oceans, lakes, rivers, and other such surface water bodies as well as groundwater. It contains 11.188 percent hydrogen and 88.812 percent oxygen by weight. It may exist as a solid, liquid, or gas. As normally found in the lithosphere, hydrosphere, and atmosphere, it may have other solid, gaseous, or liquid materials in solution or suspension. (2) To wet, supply, or irrigate with water.

Water quality The chemical, physical, and biological characteristics of water with respect to its suitability for a particular purpose. The same water may be acceptable for one purpose or use but unacceptable for another, depending on its characteristics and the requirements for the particular use. *See also* drinking water regulation; water quality criteria.

Water quality criteria Scientific standards on which a decision or judgment may be based concerning the suitability of water of a specific quality to support a designated use.

Water supply (1) In general, the sources of water for public or private uses. (2) The furnishing of a good quality potable water under satisfactory pressure for domestic, commercial, industrial, and public service, as well as an adequate quantity of water under reasonable pressure for fire fighting.

Water supply source A river, brook, stream, lake, reservoir, impoundment, spring, or aquifer from which a supply of water is or can be obtained. The use of the term *raw water* for source water is discouraged.

Water supply system (1) Collectively, all property involved in a water utility, including land, water source, collection systems, dams and hydraulic structures, water lines and appurtenances, pumping system, treatment works, and general properties. (2) In plumbing, the water distribution system in a building or complex of buildings, including appurtenances.

Water system Collectively, all property involved in the *operation* of a water utility, including land, water lines and appurtenances, pumping stations, treatment plants, and general property.

Water table The surface in an unconfined aquifer where the water pressure is atmospheric. The water table is determined by measuring the water level in shallow wells installed a few feet into the zone of saturation.

Water utility (preferred over water works) The collective set of features of a system that provides potable water to the public, including the source water facilities, treatment plant, and water distribution network. This term is typically used to designate all aspects of a given system.

Water year A continuous 12-month period during which a complete annual cycle occurs, arbitrarily selected from the presentation of data relative to hydrologic or meteorologic phenomena. The US Geological Survey uses the period October 1 to September 30 in the publication of its records of stream flow. A water year is also called a climatic year.

Watershed (1) The drainage basin area contained within in the bounds specified by a divide and above a specified point of a stream. In water supply engineering, a watershed is also called a catchment area; in river control engineering, it is called a drainage area, a drainage basin, or a catchment area. (2) The divide between drainage basins.

Well (1) An artificial excavation that derives water from the interstices of the rocks or soil that the excavation penetrates. (2) A shaft or hole into which water may be conducted by ditches to drain water away from a construction are

ABBREVIATIONS

ADEM	Alabama Department of Environmental Management
ASR	analytical services requet
ASTM	American Society for Testing and Materials
AWWA	American Water Works Association
AwwaRF	Awwa Research Foundation
BTEX	benzene, toluene, ethylbenzene, and xylene
CAA	Clean Air Act
CAL	calibration solution
CAS	Chemical Abstracts Service
CMSA	consolidated metropolitan statistical area
COB	carryover blank
CWS	community water system
CCL	candidate contaminant list
DIPE	diisopropyl ether
DMS	degree-minute-second
DNR	Department of Natural Resources (Wisconsin)
DWCCL	drinking water candidate contaminant list
DWRPD	Drinking Water and Radiological Protection Division
EPA	Environmental Protection Agency
ERA	Environmental Resources Associates
ETBE	ethyl tertiary-butyl ether
FD	field duplicate
FEB	field equipment blank
FRB	field reagent blank
FY	fiscal year
GCMS	gas chromatograph/mass spectrometer
GIS	geographic information system
GRN	Groundwater Retrieval Network (WI)
GPS	global positioning system
HA	health advisory
HCl	hydrochloric acid
HP&T-GC/MS	heated purge and trap gas chromatograph/ mass spectrometer
IB	instrument blank
IS	internal standards
KDHE	Kansas Department of Health and Environment

LDS	lowest daily calibration standard
LIMS	laboratory information management system
LRB	laboratory reagent blank
LTMDL	long term method detection limit
LUFT	leaking underground fuel tank
LUST	leaking underground storage tank
MCL	maximum contaminant level
MDEQ	Michigan Department of Environmental Quality
MDL	method detection limit
MDNR	Missouri Department of Natural Resources
MEK	methyl ethyl ketone
MRL	minimum reporting level
MSA	metropolitan statistical area
MTBE	methyl tertiary-butyl ether
MWDSC	Metropolitan Water District of Southern California
NAWQA	National Water Quality Assessment
ND	not detected
NDV	non detect value
NECMA	New England County Metropolitan Area
NIPER	National Institute for Petroleum and Energy Research
NWQL	National Water Quality Laboratory
OHSU	Oregon Health and Science University
OXY	oxygenated gasoline
PAC	Project Advisory Committee
PE	performance evaluation
PWS	public water system
PWSID	public water system identifier
RFG	reformulated gasoline
RSD	relative standard deviation
RPD	relative percent difference
RT	retention time
QA/QC	quality assurance/quality control
QC	quality control
SA	surrogate analyte
SB	system blank
SDWIS	Safe Drinking Water Information System
SOP	standard operating procedure

TAA	tertiary amyl alcohol
TAME	tertiary amyl methyl ether
TBA	tertiary butyl alcohol
TBF	tertiary butyl formate
TCE	trichloroethene
THM	trihalomethane
T&O	taste and odor
TRB	trip reagent blank
TTHM	total trihalomethanes
TRI	toxic release inventory
USEPA	United States Environmental Protection Agency
USGS	United States Geological Survey
VBW	volatile blank water
VOC	volatile organic compound
WS	water supply